MW00480134

IS THERE EVIDENCE FOR GOD?

IS THERE EVIDENCE FOR GOD?

...an Economist Searches for Answers

ROBERT GENETSKI, PH.D.

Is There Evidence for God? An Economist Searches for Answers

For information about this title or to order other books and/or electronic media, contact the publisher:

ClassicalPrinciples
Classicalprinciples.com
rgenetski@classicalprinciples.com

ISBNs:
978-0-9982227-0-7 (hardcover)
978-0-9982227-2-1 (softcover)
978-0-9982227-1-4 (eBook)
978-0-99822271-3-8 (audiobook)

Printed in the United States of America

This book is dedicated to my children, grandchildren,

and all those searching for answers about

God and the big questions in life.

CONTENTS

ACKNOWLEDGMENTS

I *would like to acknowledge* those who went through earlier versions of the manuscript, making insightful suggestions to simplify and clarify a number of issues. Along with my wife Maureen, my close friends Chad Bassett and Gary Meyers offered encouragement and suggestions to help bring the project to its current state. I alone take sole responsibility for the final product.

INTRODUCTION

I've lived a long, wonderful, fulfilling life. Although blessed with excellent health, my chronological clock tells me, in this game of life, it's late in the fourth quarter. Looking back, I have very few regrets. The main one is my failure to search for God when I was much younger, when answers could have helped me deal far more effectively with the ongoing challenges in life.

As far as questions about God and religion go, I took the easy way out. I accepted what people told me about God without seriously looking into it.

I was born into a Catholic family. Mom regularly attended Sunday Mass, and she told me about God and how important it was to always do the right thing. As with most men, Dad showed no interest in God or religion. When I was young, he seldom if ever went to church.

For most of my life, I accepted Catholic answers about God. Before going off to college, I was aware of how most students leaving for college also leave their religion behind. I recall asking God not to let it happen to me. God answered my prayer. Although I wasn't a very good Catholic in college, after marrying my wife, Maureen, we continued to make it to Sunday Mass as often as possible. I even tried to tell my children about God. However, with only a superficial knowledge of religion, my attempts to instruct others about God were seldom successful.

I always intended to look into questions about God and religion. It was one of those things on the back burner. On one occasion, I forced myself to read the entire Bible in a year. After finishing, I had no idea what I had read. Nothing. My entire knowledge of God and religion remained superficial, based entirely on what others had told me.

The limited nature of my knowledge of God became apparent during a chat with my Mom. At the time, she was close to ninety. Up to that time, she continued to go through the ritual of attending Sunday Mass. Then, several years before she would die, she confided in me that she didn't think God was real or that there was anything after death.

I was stunned! The person who first told me about God now said she didn't think he was real. It turned out a friend told her God was merely a crutch, invented for weak-minded people to help them deal with life. This challenge to her faith was something she never considered, and she was at a loss over how to confront it. And I wasn't much help.

The incident may have been the spark I needed to begin taking a serious look into God and religion. Simply accepting what others tell us about God has serious shortcomings. Had my parents and teachers had different ideas about God and religion, I'm certain I would have grown up with entirely different views.

When we accept what others tell us about God, our beliefs rest on thin ice, easily shattered by some crisis or persuasive challenge. I decided to seek answers to questions I always wanted to look into, but never seemed to have the time.

But, where to start? How would I even begin a project that had baffled mankind since the beginning of time?

My first thought was to turn to people who claim to know about God. How did they know?

Some said they instinctively know God exists. Others found God through prayer, philosophy, or through reading holy books. Some say they know God exists because he called them. Others claim to find God through science, while still others say science convinces them God isn't real.

My approach to answering questions about God would be different. In my career as an economist, I was constantly seeking evidence to complicated questions about how the world works. As with most researchers, my approach involved sifting through different views on a vast number of different topics. The tools involved digging deeply into history, philosophy, logic, science, and human nature for answers. I intended to use the research skills I had developed over more than half a century to evaluate the evidence for God.

My approach to evidence was going to be logical and systematic. I would approach the project as Mr. Spock, the fictional character from *Star Trek,* who would meticulously examine evidence, weigh the odds, and reach the logical conclusion.

The more I thought about looking into the questions surrounding God, the more I thought of the consequences. During my career as an economist, I always stressed the importance of not allowing biases to influence the analysis. I had been particularly critical of researchers who limited their research to data that supported their personal views. If I undertook the project, there would be two requirements—the analysis would have to be unbiased, and I would have to accept the evidence wherever it led.

Depending on what I found, there could be a hornet's nest of potential problems. What if the evidence failed to support my Catholic religion? What if evidence showed Christianity was a fraud? Was I prepared to change to some other religion? And, what if the evidence indicated God didn't exist? Would I be able to accept such a conclusion? What impact would my analysis have on my relationship to my family, friends, and others?

Did I really want to undertake a project with serious—even life-changing—implications?

Was such a project even feasible?

Without clear answers to any of these questions, I decided to go ahead. For better or worse, I would live with the consequences. For my own benefit, I wanted to know the truth about God.

One reason the project could be feasible was the main object of the search—God. If he existed and cared about us, he would probably want us to find him. And, if the evidence indicated he didn't exist, at least I would be true to myself in searching for him.

As the project evolved, it ended up consisting of five parts.

Part one considers the eternal God question—*To Believe or Not Believe?* Why seek to find God? What are the potential benefits in finding him? Why are some so certain God doesn't exist? Could they be right? What are the implications if they are right?

Part two deals with the supernatural. Is there any evidence of a world beyond our physical world? Is there any evidence for life after death? Do miracles occur?

Part three turns to religions of the world to determine if they provide any evidence for and about God. The search for evidence would have to be about more than just knowing God exists. If God exists, but doesn't care about us, it might be of interest to some, but irrelevant for most of us.

Part four deals with evidence for God in modern times.

The final part consists of reflections on how the evidence changed my life.

At each step, I present and evaluate evidence for and against God. I raise questions but leave it to others to determine the strengths or weaknesses of the evidence. For those who want to delve deeper into evaluating the evidence, there are detailed chapter references at the end of the book.

My project began many years ago. It began slowly before turning into an extended journey.

As the project took form, it became apparent I had jumped on an emotional roller coaster. The torrent of new ideas and information sent me scurrying in different directions for deeper insights. At times, discoveries would shake the foundations of any hope I ever had that God existed or that holy books were valid. The deeper I dug, the more questions I had, and the more I realized how inadequate my knowledge of God was, and how limited it would always have to be.

The quest had all the sensations of an Indiana Jones movie—frightening discoveries, sleepless nights, and fulfilling Ah-ha! moments with background music. There were unpleasant moments when I was terribly discouraged. There were times when I wanted to throw in the towel, took time off and even considered quitting.

This book takes you through my search for answers about God. If you decide to come along, I have to warn you—a quest for evidence of God can produce some disturbing discoveries. If you're anything like me, there will be times you'll wish you never started, times you question long-held beliefs, times you want to quit, and times you'll be glad you didn't.

I invite you to join me in the journey of a lifetime, my search for evidence of God.

PART I

To Believe or Not to Believe

Christianity asserts that every individual human being is going to live forever, and this must be either true or false. Now there are a good many things which would not be worth bothering about if I were going to live only seventy years, but which I had better bother about very seriously if I am going to live forever.

—C.S. LEWIS

Chapter 1

WHY SEARCH FOR GOD?

D*uring our lives, we make any number* of important decisions. Those decisions determine who we are and who we will become. It's often helpful to take stock of where we are in this journey of life. For better or worse, our current situation is the result of key decisions we have made. Where we will be in the future depends on important decisions we have yet to make.

What we decide about God is the most fundamental decision we will ever make. This decision is fundamental because it influences *all* other key decisions.

What we decide about God determines who we choose to befriend, what type of vocations we choose, whether we marry, who we marry, and our spousal relationship. If we have children, it will determine what we teach them as well as our relationship to them. What we decide about God will provide insights to ourselves. It will influence our outlook on life and death. In short, our decision about God greatly influences the type of life we lead and the type of person we become.

God doesn't force us to believe in him, but he does force us to make a decision about him. And, it seems there are only four choices we can make.

Conclude God doesn't exist.

Conclude we're uncertain about God, which often means ignoring him.

Claim we believe in God, and then mostly ignore him.

Conclude God exists and spend time acknowledging him and finding out more about him.

Since finding God takes time and effort, we might wonder: Why search for God?

What's in it for me?

Those who claim to know God tell us they receive wonderful benefits. They claim to know God is to know a feeling of love so intense, so deep, and so perfect, it will never abandon us. Our lives become more concerned about the well-being of others instead of satisfying our own personal desires. We will be free from concerns of what others might think about us, free from the depression, hatred, and despair that affects so many. Knowing God means no longer having any fear of death, either our own or those of our loved ones. While knowing God doesn't end pain and suffering, it can transform pain and suffering into something fruitful and productive.

How is it even possible to transform tragedies, such as the death of loved ones, into something fruitful and productive?

The answers to these and other mysteries assume God not only exists but that he cares for us. A growing number of people, particularly the youth, do not believe in God. If they're correct, spending the time and effort trying to find out about God is a waste of time.

Before delving into potential mysteries surrounding God, the first logical step is to try to determine if God is real. If God doesn't exist, it's a waste of time to ponder any details about the possible mysteries surrounding him.

The logical place to begin considering God's existence is to examine the views of those who are most convinced God doesn't exist.

Why do so many believe there is no God? Why do their views appeal to so many? What are the implications if they're correct? Are they correct?

Let's find out.

Chapter 2

THE ATHEIST'S VIEW OF GOD

A*theists claim there is no such thing as God.* They compare believing in God to believing in Santa Claus or in the tooth fairy. They claim you have to be a fool to believe in God.

To investigate atheism, I turned to three of the most dedicated atheists—Richard Dawkins, Christopher Hitchens, and Sam Harris. These atheists have studied science, history, archeology, the Bible, and the Koran, searching for answers about God. Their research leads them to conclude: There is no God. The universe, earth, and mankind are all the result of an accident of nature. We are born. We live. We die. There is nothing after death.

If God exists, these atheists have at least one thing going for them. They have likely spent more time and effort thinking and researching questions about God than most of us. Not only do they reject God: they seem to have an intense hatred for him.

Dawkins tortures his thesaurus to describe his feelings about God.

> *The God of the Old Testament is arguably the most unpleasant character in all fiction: jealous and proud of it; a petty, unjust, unforgiving control-freak; a vindictive, bloodthirsty ethnic cleanser; a misogynistic, homophobic racist, infanticidal, genocidal, filicidal, pestilential, megalomaniacal, sadomasochistic, capriciously malevolent bully. (Dawkins, 2009)*

It seems a bit odd for anyone passionately to hate something they don't believe exists. It would be like hating Santa or the Easter Bunny. Who does that?

Atheists' anger stems from their view that God and religion are the main reason for anger, hate, conflicts, wars, child abuse, depression, and just about anything else we can think of that's bad.

Could these awful things atheists say about God, religion, and the Bible be true? A quick check of Biblical references seems to show the atheists' claims appeared valid. If this nonexistent God was as bad as the atheists claim, we might be just as well off if he didn't exist.

With their passion and commitment, atheists can be very persuasive. As I immersed myself deeper into their atheist world, God and religion appeared to be deeply flawed. How could anyone possibly believe in the God they were describing? Their world of disbelief left me wondering if I had been misled as a child into believing a fairy tale about God.

After devouring their case for atheism, I began to realize how unprepared I was to deal with many of their arguments. The immense magnitude of the task ahead of me was becoming apparent. In order to fully evaluate the case against God, I would need to explore issues surrounding science, history, archaeology, philosophy, and theology.

Before undertaking such a task, there were several questions worth considering that I did feel capable of exploring.

Why is atheism so appealing, particularly to younger people?

What are the implications of atheism?

Are atheists' logic and reasons for rejecting God reliable?

Chapter 3

WHY IS ATHEISM SO APPEALING?

There are many reasons atheism appeals to so many, particularly younger people. Polls show half the children born since 1999 say they do not believe in God. If atheism were a religion, it would be the fastest-growing religion in the world.

One of atheism's most important appeals is how it can empower us and feed our egos. When we reject God, we get to take his place. We become the highest existing lifeform, beholden and subservient to no one.

Without God, there are no absolutes—no laws, no rules, no truth. Atheists are free to decide for themselves what is right and what is wrong. The only rules, laws, or truths are those they choose to impose on themselves. Who doesn't want to rule the world?

Many believe denying God gives them more freedom and control over their lives. They don't have to look over their shoulder at someone advising them how to behave or why they shouldn't do certain things. The concept of such freedom can be particularly attractive to children and young adults seeking independence from parental authority.

Children, as well as some adults, tend to associate freedom with doing whatever makes them feel good. This concept of freedom is similar to that enjoyed by a wild animal, doing whatever its nature tells it to do—roam at will, kill for food, eat, mate, sleep, and eventually die.

The difference between animals and humans is that humans have intellect. We have the freedom to reason and to control our natural instincts. When we fail to control them, our behavior descends to that of a lower form of life, as well as a lower concept of freedom.

Human nature is such that those who choose to define their own rules often tend to choose things that produce immediate satisfaction. These include bingeing on food, Netflix, video games, sports, alcohol, drugs, and sex, or immersing oneself into the artificial life of the metaverse. Satisfying such freedoms too often leads to a desire for progressively more intense personal satisfaction. Too often, such pursuits lead to destructive behavior, to a life of addiction, depression, a life without purpose. Ironically, the freedom from rejecting God and his guidance can lead to having less, rather than more, control over your life.

A higher concept of freedom occurs when we control our instincts, rather than have our instincts control us. Some people are able to control their basic instincts and adopt high moral standards without believing in God. However, a belief in God and his guidelines can make it easier to control our instincts and live a more purposeful and productive life, something that's easier to do when an individual is living for eternity, instead of for the moment.

One clear distinction between those who believe in God and those who don't involves freedom from the fear of death. Among the greatest tragedies we experience in our lives is the death of our loved ones. What are the implications of atheism for those emotionally trying times in our lives when we face our death or the death of our loved ones?

Chapter 4

THE IMPLICATIONS OF ATHEISM

As a child, everyone I knew believed in God. So, I believed in him. I believed God created us and all things; our purpose was to know, love, and serve him. When we die, we would share eternity with him and our loved ones. Whether my childhood views were correct or not, they gave me a very positive outlook on both life and death.

The atheist outlook is different. Without God, the universe, Earth, and all of us are an accident of nature. There is no reason we were born and no reason for us to live, and, when we die, it all ends. We and our loved ones are gone forever.

Atheists who have thought deeply about these implications tell us about the impact their philosophy has on their lives. David Sinclair is a brilliant scientist at the cutting edge of using genetics to help reverse aging. In his book *Lifespan,* he provides his personal perspective on the atheist view of death:

Even if they don't recognize its violence, children come to understand the tragedy of death surprisingly early in their lives. By the age of four or five they know death occurs and is irreversible. It is a shocking thought for them, a nightmare that is real.

. . . . Between five and seven, however, all children come to understand the universality of death. Every family member will die. Every pet. Everything they love. Themselves, too. I can remember first learning this. I can also very well remember our oldest child, Alex, learning it.

"Dad, you won't always be around?"

"Sadly, no," I said.

Alex cried on and off for a few days; then he stopped and never asked me about it again. And I never mentioned it, either.

It doesn't take long for the tragic thought to be buried deep in the recesses of our subconscious. When asked if they worry about death, children tend to say that they don't think about it. If asked what they do think about it, they say it is not a concern because it will occur only in the remote future, when they get old.

That's a view most of us maintain until well into our fifties. Death is simply too sad and paralyzing to dwell on each day. Often, we realize it too late. When it comes knocking, and we are not prepared, it can be devastating.

It takes courage to consciously think about your loved ones' mortality before it actually happens. It takes even more courage to deeply ponder your own. (Sinclair, 2019)

Sam Harris presents his view on the atheist's outlook on life:

> *This life, when surveyed with a broad glance, presents little more than a vast spectacle of loss. . . . When the stopper on this life is pulled by an unseen hand, there will have been, in the final reckoning, no acquisition of anything at all.*
>
> *And as if this were not insult enough, most of us suffer the quiet discomposure, if not frank unhappiness, of our neuroses in the meantime. . . . We are terrified of our creaturely insignificance, and much of what we do with our lives is a rather transparent attempt to keep this fear at bay. . . . nearly the only thing we can be certain of in this life is that we will one day die and leave everything behind. . . . (Harris, 2004)*

It's difficult to imagine a more depressing outlook on life. By their own admission, atheists live a life filled with neuroses surrounding their fear of death.

Atheists' difficulty in dealing with death stands in stark contrast to my childhood view. I knew about death. It never bothered me. Assuming we all behaved, I expected my loved ones and I would all go to Heaven to enjoy eternity.

At various times, I would even think about my last day on Earth. Would I have time to review the events of my life? Would I be pleased about how I lived and the decisions I made? Would I lie there with regrets over what I had done?

Most important of all, when the end came, would I be at peace in the belief that I had lived as God wanted me to live?

I'm not sure this prospective, retrogressive approach to death can be helpful to others. It was helpful to me to think about how to live while there is still time to do something about it, rather than wait until it's too late to make any meaningful changes.

Even if atheists are correct and I was misled about God and an afterlife, even if I was wrong to trust in God and the next life, I am still thankful for my childhood perspective. Whether atheists are right or wrong, I'm grateful my immature childhood mind did not have to cope with the atheist's implications for life and death.

Some atheists recognize the depressing implications of their views and keep them to themselves. They let others enjoy whatever hope and comfort they might get from believing in God, even if they don't believe it themselves. Although they may believe God is a fantasy, they know it can be a useful one.

Whether atheists' conclusions about God are right or wrong, there is one thing atheists appear to have done—they appear to have managed to create their own Hell here on Earth.

Chapter 5

NO LIVES MATTER

T*he following is a satirical news item* from Babylon Bee (a news and satire website). It captures the philosophical implications of the atheist's world.

ATHEISTS LAUNCH NO LIVES MATTER MOVEMENT

WORLD—A group of atheists, along with some agnostics, announced on Tuesday a new sociopolitical movement consistent with their worldview called No Lives Matter.

According to sources, organizers for NLM have planned numerous rallies to protest other rallies claiming that lives matter. The organization's mission statement defines the group as "people motivated by the belief that all human lives are equally meaningless."

"Since we are just random accidents of evolution, and our so-called moral truths are just biochemical reactions

in our brain, no human lives, or any lives at all, actually matter," said prominent atheist Richard Dawkins, speaking for NLM. "Anyone who says Black Lives Matter, or Blue Lives, or whatever lives, has to believe in a higher moral code given by a creator. You can't say in one breath that lives matter, and in the next breath claim that God is a human invention and all truth is relative."

"We're just trying to be logically consistent," he added.

The movement quickly fell apart, though, when leaders had trouble convincing followers that the No Lives Matter movement mattered.

As depressing as the implications of atheism can be, we should never reject a conclusion because it's unpleasant or because we might not like its implications. Most of us are hardwired to seek truth. Atheists say they have found it.

Let's evaluate the case for atheism and determine if it's sufficiently persuasive to reject the idea of God.

Chapter 6

EVALUATING THE CASE FOR ATHEISM

At *this stage* I wasn't sufficiently knowledgeable about many of the fields atheists discussed. That would have to come later. At this point, I was more concerned with the logic, critical reasoning, and evidence atheists used to reach their conclusion.

Reading past atheists' emotion and passion against God, their case for rejecting him relies on the following five pillars. God does not exist because:

There is no evidence God exists

God is inconsistent with science

God wouldn't allow bad things to happen

Religious people do bad things

Intelligent people know God doesn't exist

1. There is no evidence God exists. This is the primary pillar for atheism. Each of the dedicated atheists incessantly tell us there is no evidence for God. In his book, *The End of Faith,* Sam Harris mentions the lack of evidence for God on almost every other page. On one page he tells his readers *four times* there is no evidence!

This main pillar is simply wrong. There is evidence for almost every point of view on almost every issue imaginable. Our task is to examine the evidence and determine its strengths and weaknesses. To make their case effectively, atheists should present the evidence for God fully and then give their reasons for rejecting it. They fail to do so.

After repeatedly telling us there is no evidence for God, Dawkins goes on to mention evidence. He refers to the miracle at Fatima:

> *On the face of it, mass visions, such as the report that seventy thousand pilgrims at Fatima in Portugal in 1917 saw the sun "tear itself from the heavens and come crashing down upon the multitude," are harder to write off. It is not easy to explain how seventy thousand people could share the same hallucination. But it is even harder to accept that it really happened without the rest of the world, outside Fatima, seeing it too—and not just seeing it, but feeling it as the catastrophic destruction of the solar system, including acceleration forces sufficient to hurl everybody into space. (Dawkins, 2009)*

After acknowledging how a miracle that was seen by 70,000 people is difficult to dismiss, Dawkins dismisses it. He tells us it couldn't have happened because God and supernatural events cannot occur.

Hitchens also mentions evidence surrounding the *Shroud of Turin,* the assumed burial clothes of Jesus. A quick Google search indicates there has been more than half a century of detailed scientific investigations into the *Shroud.* Investigations into its authenticity have been the most extensive and detailed in the history of any relic. Rather than present and discuss the strengths and weaknesses of the scientific evidence, Hitchens dismisses it in four words by referring to the "discredited Shroud of Turin." (Hitchens, 2009).

The reason atheists are so quick to dismiss evidence for God appears to be directly related to their belief that God isn't real. Sam Harris captures their position, explaining, "*. . . every religion preaches the truth of propositions for which it has no evidence. In fact, every religion preaches the truth of propositions for which no evidence is even conceivable. (Harris, 2004).*

Imagine you are on trial for a crime you didn't commit. How confident could you be if the jury couldn't *even conceive* of your innocence? By their own admission, atheists *can't even conceive of the possibility of God.* Hence, they refuse to consider, much less evaluate, any evidence pointing to God's existence.

2. God's existence is inconsistent with science. If God exists, and if he created the universe and all of mankind, he is clearly omnipotent, omniscient, and beyond time and space. By his very nature, God has to be inconsistent with our concept of science. To create the physical world, God had to be beyond our physical scientific world. To claim God doesn't exist because he's inconsistent with science isn't a logical reason for believing he doesn't exist. You can't logically start with the assumption God doesn't exist, and then use this assumption as the reason he couldn't exist. It's faulty logic.

3. God wouldn't allow bad things to happen. Atheists tell us God doesn't exist because atrocities occur. They apparently believe if God exists, he was supposed to have created a paradise on Earth. Maybe God did create a paradise on Earth. Maybe we didn't deserve it. Maybe he should have given us a paradise whether we deserve it or not. Maybe God prefers to test us to see how we deal with the evil, pain, and suffering of this world, and then determine who belongs in his paradise. Assuming God should have behaved in one way or another without offering any evidence is simply offering a personal opinion. Personal opinions are not substantive reasons to either accept or reject any conclusion.

4. Bad things have been done by people claiming to be religious. Dedicated atheists begin by telling us: you can't blame a person's religion for the person's personal behavior. Then, they do what they tell us not to do. They go through centuries of history blaming religions based on love and forgiveness for the behavior of members who fail to be loving and forgiving. The contradiction in their reasoning undermines its validity.

5. *Intelligent people don't believe in God.* Dawkins, Hitchens, and Harris all tell us they are extremely intelligent. One reason they give for their intelligence is that they don't believe in God. They make it clear that, if you believe in God, you aren't very intelligent.

Harris tells us intellectuals know that reason is supreme over faith: ". . . intellectuals . . . have declared the war between reason and faith to be long over." (Harris, 2004)

Dawkins praises those ". . . whose native intelligence is strong enough to overcome . . ." a belief in God. (Dawkins, 2009)

Dawkins also quotes studies showing ". . . the higher one's intelligence or educational level, the less one is likely to be religious, or hold 'beliefs' of any kind." (Dawkins, 2009)

Hitchens' vain attempt to conceal his intelligence is even less subtle. "It might be immodest to suggest that the odds rather favor the intelligence and curiosity of the atheists. . . . " (Hitchens, 2009).

Hence, the atheists' final reason not to believe in God is to claim if you don't agree with them, you aren't very intelligent.

Of all the reasons atheists give for not believing in God, the most substantive is their claim there is no evidence. However, when atheists tell us they cannot even conceive of God's existence, or of any evidence for his existence, they make it clear they are not interested in evidence. Without an inclination to examine evidence, the atheist case for rejecting God rests entirely on their personal opinions and beliefs.

Atheists assume their beliefs reflect reality. However, reality is one of two things—God either exists, or he doesn't. Reality isn't determined by personal beliefs. Reality is based on evidence. The failure of atheists even to conceive of evidence is a serious shortcoming to accepting their conclusion. The Bible tells us, *All who seek shall find, and to all who knock the door will be opened.*

Atheists refuse to seek because there is nothing to find. They will not knock, because there is no door.

After initially believing that atheists might have good reasons for believing God does not exist, further evaluation led me to reconsider. Their initial bias against God's existence was so great, they appeared blind to any evidence to the contrary.

None of these shortcomings is sufficient to reject the atheist's conclusion. God may still not exist in spite of the many problems with the atheist's reasoning. After considering the substance of the atheist case against God, my main conclusion is that they are not going to be very helpful in seeking the truth about God.

Our quest for God continues. We will seek in an effort to find, and we will knock to see if anyone answers. Our next step is to seek evidence for life after death.

PART II

Life After Death

Never will we understand the value of time better than when our last hour is at hand.

—ST. ARNOLD JANSSEN

Chapter 7

LIFE AFTER LIFE

D*oes everything end when we die?* Or, is there evidence of another world beyond the physical world we live in? If there is evidence for life after death, it brings us a step closer to believing in God.

Until about half a century ago, scientists left questions about life after death to psychics, mediums, and religious authorities. In the 1960s Raymond Moody, MD, became fascinated by extraordinary out-of-body experiences of individuals who had been visibly dead. In 1975, he published the accounts of more than a hundred incidents in his book, *Life After Life*.

Each reported experience was somewhat unique. And yet, there were similar patterns to their stories. For example, many of the individuals described having spiritual bodies with enhanced abilities of sensory perception and mobility. Most described being immersed in a world filled with love and beauty so intense that there were no words to describe it. Some had the feeling they had been there before and were returning home.

Unlike in dreams or hallucinations, which are often hazy and confusing, the world they describe was *more real* than the one they lived in on Earth. They claimed it was as if their normal world were covered with a veil, while in this new world, the veil was lifted; for the first time they could see things with complete clarity.

Moody documents how certain cases had corroborating evidence. For example, some would describe meeting dead relatives they never knew; others told of conversations they overheard of those in hospital waiting rooms. Family members verified their stories after the individual had returned to life. In one instance, a person told of floating outside the hospital and described an item not visible from any other vantage point. An investigation revealed that the item was where the person said it was.

Moody was also fascinated with how these out-of-body experiences changed the lives of those involved. Most had a totally new perspective on life. They became more focused on loving and helping others and no longer had any fear of death. The experiences occurred irrespective of any belief in God or of any religious beliefs.

Not all those who were visibly dead had pleasant experiences. Some reported nothing. A small number told of unpleasant, even horrifying, experiences.

Moody's book opened the door to an outpouring of scientific analysis of what is now known as "near-death-experiences" (NDEs). Evidence of a world beyond our own has become progressively more common. In 2006, medical researchers from around the world met to discuss their research and present peer-reviewed analyses of their work. Their extensive research was published in 2009 in *The Handbook of Near-Death Experiences: Thirty Years of Investigation*.

These more recent comparisons of NDEs are remarkably consistent with those compiled by Moody half a century earlier. If individuals actually experienced a world beyond ours, that world hasn't changed.

Moody speculated that one out of thirty people had such an experience. In his 2009 book, *Imagine Heaven*, John Burke presents an extensive description of NDEs from books, articles, and personal interviews. He speculates one out of twenty-five people have likely had such experiences.

Scientific investigations of NDEs indicate they are neither dreams nor hallucinations. The events are so real that they lead to lasting changes in people's lives. People become more sensitive and responsible for the well-being of others and have a healthy perspective on what happens after we die.

What are we to make of such experiences? Are they simply the product of vivid imaginations? Or do the large number of cases and the similarity of many of the experiences suggest we should take them seriously?

When events the person couldn't know about (such as knowledge of dead relatives) are confirmed by others, it would seem to provide some evidence of a world beyond ours. Finally, the dramatic changes that occur in people's lives after NDEs, such as committed atheists suddenly becoming believers in God, seem to provide persuasive evidence for such a world. At the very least, it provides strong evidence for certain former atheists.

Before evaluating the evidence for a world beyond ours, it can be helpful to consider the stories of three people whose NDEs had a remarkable effect on their lives, as well as on the lives of many others.

Chapter 8

VISITS TO THE WORLD BEYOND

A*t about 9 p.m. on a chilling evening* in March 1999, Tommy Rosa went for an evening walk. As he headed back to his apartment in the Bronx, his world was about to change. Suddenly, a speeding car without lights sent him hurtling into the air. His body skidded on the asphalt, smashing skin and muscle against his bones. He was unconscious, died for several minutes, and then was resuscitated. He spent weeks in a coma before an extended period of rehabilitation enabled him to return to his job as a plumber.

As with many who have had NDEs, he was reluctant to talk about it. A chance meeting with a medical doctor led to a friendship and talks about Tommy's experience. Dr. Stephen Sinatra had already encountered the phenomenon. He explains,

> *I'm no stranger to the NDE phenomenon. For many years, I worked in emergency rooms, coronary and intensive care units, and cardiac catheterization labs where patients were suddenly brought back from the brink of death. I've heard nearly 20 NDE reports from heart-attack survivors who subsequently became more intuitive, grounded, giving, and loving—and less materialistic—after visiting the other side.*

Tommy's NDE was unique. He claims he was taught a number of divine lessons during his stay in what he assumed was Heaven. Many of the lessons were about how to keep healthy and avoid getting sick. Dr. Sinatra confirmed that what Tommy had learned in Heaven was entirely consistent with the most recent medical discoveries.

Their discussions led them to co-author *Health Revelations from Heaven and Earth*. The book includes all the lessons Tommy received in Heaven for maintaining good health. Dr. Sinatra then provides corresponding medical studies explaining why these recommendations are so effective. How does a plumber, with no knowledge of medicine, somehow end up with confirmed insights to the latest medical research?

What would it take for an atheist, who couldn't *even conceive* of the existence of God, to become a believer? Trips to Heaven can be very effective, as former atheists have discovered. The following two conversions appear particularly interesting.

Eben Alexander spent 15 years on the faculty of Harvard Medical School, specializing in neurosurgery. While there, he operated on countless patients, many with severe, life-threatening conditions. He authored or co-authored more than 150 chapters in peer-reviewed medical journals and presented his findings at more than 200 conferences around the world. Dr. Alexander had spent his whole life healing. As with many doctors and other scientists, he did not believe in God and was convinced life after death was impossible.

On Monday, November 10, 2008, Dr. Alexander woke up to a sharp back pain. His wife found him unconscious. He was in a coma that would last seven days. During this time, the neocortex of his brain, the part that makes us human, was shut down. He wasn't clinically dead. However, without any function in his neocortex, he could not generate conscious thoughts. His body functioned in a vegetative state, as near to death as possible. A ventilator and feeding tube kept him alive. On Sunday morning of the seventh day, with all hope seemingly gone, his doctors recommended the removal of all life support.

During his coma, Dr. Alexander claims he first had to go through a dark tunnel, a not-so-pleasant trip. He then experienced a transformation to a miraculous dimension of love and peace. He describes the experience as so real, and so magnificent, words are incapable of describing it.

While each reported NDE differs in some ways, Alexander's experience is unique for two reasons. As an expert in neurosurgery, his expertise gave him a professional perspective on the brain, how it functions, and what it is capable of doing. Also, his NDE, without a functioning neocortex, lasted far longer than any others. As a result, he says he went much deeper into his NDE than others he has since read about. The fascinating description of his deep dive into another world is in his book, *Proof of Heaven*.

Roy Schoeman was raised Jewish. He expected that his bar mitzvah would enable him to have a personal relationship with God. When nothing magical happened, he says it was the saddest day in his life. In college, he abandoned both his Jewish religion and his belief in God. As with many of his Harvard colleagues, he viewed religion as a medieval superstition, one science had made obsolete.

At the age of 29, Schoeman achieved his wildest dream. He was chosen to be a Professor of Marketing at the Harvard Business School. In spite of achieving the ultimate in worldly recognition and success at such a young age, Schoeman fell into a state of utter despair. As with so many atheists, he believed we and our existence were all an accident of nature. Just as others who had grasped the full implications of such a conclusion, he felt there was no meaning or purpose to life. In a severe state of depression, he contemplated suicide.

One morning, as he was walking along a beautiful nature preserve in Cape Cod, Schoeman says, "The veil between Heaven and Earth disappeared, and I found myself in the presence of God, seeing my life as I would see it after I died, and was looking over my life in the presence of God."

Schoeman says he was overwhelmed with an ocean of love from an all-loving God who had loved him all his life as if he were the only person in the world.

"I saw that it was all true, that we live forever, that every action has a moral content, that it is observed, recorded, and weighed in the balance and recorded for all eternity. That every right decision we make, every right moral action, will be benefiting us for all eternity, and every lost opportunity, when we do not do something righteous, will be a lost opportunity for all eternity."

Schoeman says he experienced true reality. He recognized the world we live in as something less than reality. The experience changed his life forever. He is no longer an atheist; he has become an evangelist. As a researcher, his website contains a number of his speeches and accounts of his beliefs. His fascinating videos of an experience with God and Heaven are readily available for those who wish to evaluate both his insights and his credibility.

A plumber receives valuable insights to recent medical advances. Committed atheists become evangelists. Clearly, something powerful is happening. For an estimated 4% of people who have such experiences, the evidence is clear and compelling—there is life after death. God not only exists—he allows certain people a glimpse of what awaits them in the world to come.

For the 96% of us who haven't had such experiences, we have to evaluate the evidence secondhand, assess the credibility of the witnesses, and decide to accept or reject their accounts. We might also wonder about those out-of-this-world experiences that are unpleasant. It would seem they point to the potential for two very different worlds. For some, a world eternally filled with love and peace. For others, a world where the physics of cosmic justice provides an alternative to eternal peace. Unfortunately, those with unpleasant experiences tend to be reluctant to provide details.

We'll conclude this section by examining the evidence for miracles and what they might be able to tell us about a world beyond the physical one we live in.

Chapter 9

IT'S A MIRACLE!

"There are only two ways to live your life. One is as though nothing is a miracle. The other is as if everything is."

—ALBERT EINSTEIN

A *miracle is said to occur* when God suspends the laws of nature to increase our faith and knowledge of him. Miracles assume God is real. Since atheists believe there is no such thing as God, they also believe there is no such thing as a miracle.

Professor Charles Keener provides a detailed, comprehensive analysis of miracles in his two-volume book, appropriately titled, *Miracles*. Keener's research shows how almost all societies, ancient and modern, have experienced miracles. He explains how, in the modern world, miracles occur more frequently in third-world countries, where healthcare is not readily available, than in highly developed countries.

Keener describes how Pentecostal, charismatic Protestant, and charismatic Catholic churches all report high levels of faith healing after prayer. These healings have led to such rapid growth that the charismatic branch of Christendom is second in size only to Roman Catholicism, which also embraces supernatural charismatic beliefs.

Keener claims miracles are very common. As with near-death experiences, they occur regardless of a person's religion or lack of religion. He says an easy way to confirm the pervasiveness of miracles is to survey random groups, such as students. Doing so often reveals a large number of individuals who have personally experienced supernatural events or are closely related to someone who has. Keener relates details for such dramatic occurrences as blind people gaining sight, cripples suddenly walking, and even dead people coming back to life.

When a person experiences a miracle—which defies all medical science—it's natural to turn to God. Miracles often serve to draw people closer to God.

Do miracles occur? Is God really involved in our lives?

Many assumed miracles have completely rational explanations. The Catholic Church, which takes claims of miracles seriously, has a strict procedure for determining if a presumed miracle has likely occurred, or if an event may have a completely rational explanation.

To identify an alleged miracle, the first step is for the local bishop to establish a commission of experts. These experts include doctors, theologians, psychologists, and other scientists with specific expertise in the subject at hand. The commission is charged to operate under a suspicion of doubt. Through a detailed process, it holds extensive interviews with all those involved. The experts are expected to have a bias toward rejecting the event as a miracle and assume it has a more rational explanation.

The commission rejects most of the "miracles" it examines. For example, throughout the history of the Catholic Church, there have been numerous instances of individuals claiming Jesus' Mother Mary appeared to them. In the 20th century alone, the Church recorded 386 cases of such claims. Of these, 20% were judged as not believable, 77% were judged to have insufficient evidence. Only 2% passed the test. Even when an event passes these tests, the Church does not claim the miracle is true. It simply states the event is "worthy of belief."

Professor Keener claims this process is far too demanding and misses many true miracles.

If miracles are as common as Keener's research suggests, it's difficult to dismiss them. Those with firsthand experience are certain a miracle occurred. However, those hearing about it from others are relying on secondhand information. As with near-death experiences, their belief in the miracle will usually rest more on their faith in the individual, or in God, than on hard evidence.

Each of us can weigh the evidence for life after death. The stories and accounts of those who claim to have experienced another world or dimension are extensive. This is particularly true when there is corroborating evidence.

I was particularly impressed by the character and sincerity of those who told of their experiences. Their personal accounts are available on the Internet or through books for anyone who wishes to make their own assessment of the validity of their experiences. Looking into these stories can provide surprising benefits. For example, the book *Health Revelations from Heaven and Earth* provides a number of recommendations I continue to follow, recommendations that appear to have helped improve my health.

While looking further into life-after-life experiences, it's apparent all the evidence involves the experiences of an individual or a small group of individuals. We have to accept the word of the individual or some eyewitnesses about what happened. Each of us will then either accept or reject the stories based on our perception of the credibility of the person, or on our faith or biases regarding God. In my search, I am hoping for more details about God than the personal stories of others.

Religions are all about the supernatural. They can have a lot to say about God. My hope is the religions of the world will provide some additional evidence for God, his world, and what he's like.

Let's turn to religions and find out what they have to offer.

PART III

Religions and the Search for God

There are two ways to be fooled. One is to believe what isn't true; the other is to refuse to believe what is true.

—SØREN KIERKEGAARD

Chapter 10

HINDUISM, BUDDHISM, AND NEW AGE RELIGIONS

The *great religions of the world* fall into two broad categories. One group bases their beliefs primarily on introspection, looking inward for answers to the big questions about God and life. The other group claims their beliefs come directly from God and his messengers.

Hinduism, Buddhism, and most New Age religions base their beliefs on introspection. To find answers to questions about God and life, an individual or a religious leader looks inward, into their essence or soul. Looking inward involves meditation. Through meditation the individual expects to gain insights and answers to the big questions about life. Scientific studies show meditation and deep thinking can produce wonderful physical and emotional benefits.

For religions that rely on introspection, answers to the big questions about God and life are not found in traditional sacred or holy books. Introspection can be very appealing, particularly to those who might be suspicious of holy books or of organized religions.

Many so-called *New Age* religions have emerged in recent years. In looking inward for truth, *New Age* religions provide a wide range of constructive ideas surrounding peace, love, and how to live. Specific beliefs are usually not clearly defined; they can vary significantly from one group to another. In some cases, there is a reference to God. In other cases, the focus is more generally on love, peace, and communing with nature and the universe.

The term *New Age* implies something new, exciting, and intriguing. With religion and the search for God, there is seldom anything new. The foundations for *New Age* religions were laid 4,000 years ago with the emergence of Hinduism, and again 2,500 years ago with the emergence of Buddhism.

Hinduism's earliest religious leaders provided no clear answers about God. They believed questions surrounding the origin of the universe and God were beyond the scope of human understanding. Hence, they concluded there may be one God, or many gods, or no god.

Traditional Hindu beliefs dealt with a person's position in this life, which depends upon fate, God, gods, and on how we behaved in a previous life. Our objective in this life is to accept whatever fate or position we have been given, behave honorably, and take care of ourselves and others. Do this, and when we die, we will come back to a new and better position in our next life.

Hinduism's belief in reincarnation, of being reborn into new lives, can only end when we achieve *nirvana*—a oneness with either God or the universe. Hinduism claims we can't achieve oneness by reading books. To achieve *nirvana,* we must give up worldly pleasure and look inward.

Buddhism has many similarities to Hinduism. A major difference is its founder, Siddhartha Gautama, Buddha for short. As with us, Buddha wanted answers to the big questions in life. To get them, he sat down under a tree and spent a long time thinking deep thoughts. As with Hinduism, Buddha's answers involve the idea of reincarnation and *nirvana* as the ultimate goal in life. As with Hinduism, Buddha concludes the ultimate goal can be achieved only by meditation and looking within.

All of these religions provide an ethical formula for living a good life and seeking something beyond our physical world. Specific beliefs are not clearly defined and have evolved over time. In some cases, there is a reference to God. In other cases, the focus is directed more generally to love, peace, performing certain rituals, and communing with nature or the universe.

Can intense meditation help us to find answers about God and the great questions in life?

Perhaps.

One obvious problem in seeking answers to the big questions about God and life is how looking inward can produce a lot of different answers. The approach also raises an important question that supersedes other questions.

How do we know the answers we're getting from meditation are true? How do we know they aren't what we want the answers to be?

While inward-looking religions have attracted many followers, the wide variety of answers and lack of clarity concerning God suggest they aren't going to be much help in our search for evidence of God.

The select few who claim to have achieved *nirvana* say they have found the answers. They may be correct. If God exists, it's entirely conceivable he would reveal himself to those who spend a great deal of time and effort searching for him.

However, those who claim enlightenment aren't always clear about how their answers can help us. Some say they have met God; others claim a oneness with the universe; still others refuse to describe their experience, claiming *nirvana* must be experienced to understand it.

For those who have achieved enlightenment, it must be awesome. While they may have all the answers we're looking for, it isn't clear how their personal revelations can help us in our search for evidence for God.

Next, we turn to those religions that claim their beliefs come from God. If true, they might be helpful in our search for evidence of God.

Chapter 11

JUDAISM, CHRISTIANITY, AND ISLAM

Judaism, Christianity, and Islam all claim their beliefs come from God. All three claim God's revelations are described in sacred books—the Bible, which consists of the Old Testament and the New Testament, and the Koran.

Each religion claims God revealed the truth about himself either directly or through prophets who relayed God's messages. Moreover, they claim their sacred books contain the answers we seek to questions about God and life.

Whatever truth these books may contain, the secular world has relentlessly persecuted those closely associated with each of these religions. And, at one time or another, members of each of these religions have persecuted each other. Intense hatred and bloody wars erupted. As if conflicts among the three main religions weren't bad enough, at times conflicts within Christianity have been every bit as intense and bloody as anyone could imagine.

If God did speak to believers of these religions, what could he have said to lead to the conflicts that followed?

Searching through these sacred books reveals some interesting conclusions.

Each of these religions accepts the Old Testament as the inspired work of God. Therefore, each recognizes there is one all-powerful, all-knowing God who is beyond time and space. He is the same God, whether we choose to call him the Almighty, El, Yahweh, Lord, Father, Yeshua, Jesus, or Allah. Whatever we call him, he is the one God who created the heavens and the Earth and created each of us in his image.

All three sacred books agree on what God wants us to do—worship him, recognize the temporary nature of the material world and material possessions, behave with honesty and integrity, help others who are less fortunate than us, and live peacefully with one another.

These religions are almost unanimous in agreeing that each person has a soul whose eternal destiny depends on us doing God's will. An exception to this view exists within Judaism. The Bible describes two Jewish tribes, who disagreed about the existence of an afterlife. Interestingly, this debate over an afterlife continues within Judaism. To this day, Orthodox Jews tend to believe in the afterlife, while many secular Jews reject the idea.

All three religions agree the Jews were originally God's chosen people. They also agree that about 4,000 years ago, God chose a devout man—Abraham—to be the father of many nations. Abraham's sons, Isaac and Ishmael, became the fathers of the Jews and Arabs. Isaac's son Jacob is best known today as the patriarch from the stage musical *Joseph and the Amazing Technicolor Dreamcoat*. God changed Jacob's name to Israel, and his twelve sons became heads of the twelve tribes of Israel.

Although Jews and Arabs may be reluctant to admit it, they are brothers and sisters.

While the three religions have much in common, there are significant differences. Orthodox Jews believe God will send a *messiah,* a savior who will bring salvation to the Jewish people and to all nations. Christians believe Jesus is God's promised *messiah* who has come to save mankind. Some Jews also believe Jesus is the promised savior. They consider themselves Jewish-Christians or Messianic Jews. The Christian Bible includes both the Old Testament and the New Testament, which includes stories of Jesus' life and teachings.

Muslims believe the archangel Gabriel revealed the Koran to Muhammad about 600 years after the birth of Jesus. The Koran accepts most of the Bible as the inspired word of God. For example, the Koran refers to and praises Jesus and tells us to honor his mother, the Virgin Mary. The Koran also praises Jesus' apostles. Much of the Koran preaches peace and tolerance toward the Jews and Christians. While speaking of differences among the religions, the archangel Gabriel speaks to Muhammad using the royal *we*:

We have assigned a law and a path to each of you. If God had so willed, He would have made you one community, but He wanted to test you through that which He has given you, so race to do good: you will all return to God and He will make clear to you the matters you differed about. Sura 5:48 (M.A.S. Abdel, 2004)

As in the Bible, there are parts of the Koran that can be taken out of context to justify mistreatment or violence to others. In spite of such references, the main message from each of these books is remarkably similar—an all-powerful God created us all; he wants us to love and serve him; if we live a life of peace and help those in need, we will enjoy eternity with him in Heaven.

The most important way Christianity differs from both Islam and Judaism is with respect to God's nature. The Koran praises Jesus as a great prophet, but insists Jesus is not God. Most Orthodox Jews agree with Muslims on this point.

Christians agree with Jews and Muslims that there is only one God. However, Christians believe God's nature is different from our human nature. Christianity holds that God's nature consists of three parts—the Father, the son Jesus, and the Holy Spirit. Each of these parts is so intimately related to each other that each make up one God. While the Old Testament and the Koran explicitly mention the spirit of God, neither refers to the spirit as part of God's nature.

As creator of the universe, God's nature has to be far different from our human nature. Explaining God's nature to humans is probably as challenging as we humans trying to explain our nature to our dogs.

Differences among these three religions appear secondary to answering the big questions about God and how we should live. All three religions are part of the same story, the same God, the same relationship between God and man, and each has the same answers to life's big questions. They all have similar requirements to achieve salvation with God when our life on Earth is over.

If the instructions in these sacred books are so similar, what explains the resentment, hatred, and even wars among those who follow similar sacred books?

The simple answer is those claiming to be religious didn't know, or chose not to follow, their sacred books. Although often claiming to believe in God and accepting the beliefs of their respective religion, many Jews, Christians, and Muslims seem not to care very much about what their sacred books say, or the details surrounding their religion.

History indicates people who attain power over others often abuse such power. This is true of all people, whether religious or not. The problem associated with religious wars and violence has never been related to the religions' fundamental message. Conflicts arise when those associated with the religions fail to obey their sacred books.

Each of these sacred texts contain much wisdom—but is the wisdom from God? Or are we dealing with man-made quotes from some humans with hyperactive imaginations?

To attempt to answer this question, let's investigate where the sacred texts come from and evaluate the evidence for the claim that they might have been inspired by God.

Chapter 12

DISCOVERING THE BIBLE'S OLD TESTAMENT

T*he Old Testament, or Hebrew Bible,* provides the foundation for Judaism, Christianity, and Islam. It is an integral part of the most popular, most read, most studied, most scrutinized book ever written. It may also be the most complicated and most confusing book ever written. Although widely ridiculed, it has brought comfort and inspiration to people for more than 2,500 years.

One reason for confusion is that the stories, poems, and events were written at very different times by very different people. To make things even more complicated, most parts can be read on different levels. There are simple, basic stories, written so a child can readily understand them. For those who dig deeper, there is advice on human nature, how to behave, and how to deal with the major challenges in life. And, for those who are willing and able to dig even deeper, there are highly symbolic, theological insights into God and our relationship to him and what he wants.

Many have spent a lifetime studying and pondering the Bible, only to scratch the surface. Its popularity stems in part from its clear answers to life's big questions.

Is there a God? Yes.

What's he like? He's all powerful, he created all things, and he is beyond time and space.

Why are we here? Because he created us.

Is there life after death? Yes, our souls are eternal.

How should we live our lives? Knowing, loving, and serving God and others.

These answers provide the foundation for Judaism, Christianity, and Islam.

For many, the Bible is the *inspired* word of God. A small minority consider it the *literal* word of God. The more Biblical scholars delve into the origins of the Old Testament, the less likely it appears to be the literal word of God.

Some Biblical scholars believe they have found evidence surrounding the actual human authors who wrote the Old Testament. With the use of literary, linguistic, and archeological history, Biblical scholars have also found what they consider serious mistakes by the authors.

We humans have flaws. We make mistakes. When relating a story, we tend to tell it from our own personal experience and perspective. We sometimes use hyperbole and symbols to make a point. We may use our imagination and even embellish a story, to make it more interesting.

Scholars believe many of the sacred authors had such tendencies.

Originally, many believed the early parts of the Bible were written either by Moses or his associates. The more scholars delved into its literary, linguistic, and archeological history, the more skeptical they became. Some scholars even began to question the Bible's validity as a historical document, not to mention the inspired work of God.

In their book, *The Bible Unearthed,* Israel Finkelstein and Neil Asher Silberman conclude the archaeological history suggests the Hebrew Bible is pure fiction. They believe the story of creation, Abraham's encounter with God, the Exodus from Egypt, and most of the history of Israel presented in the Old Testament are "*. . . not a miraculous revelation, but a brilliant product of the human imagination.*"

One thing almost all Biblical scholars agree on is how different parts of the Old Testament were written at different times by different people. They believe all of the writers were likely Jewish religious leaders. Each provides a different perspective on the oral and written history that preceded them. The original stories, events, and traditions mentioned in the Hebrew Bible seem to have occurred somewhere around 1,000 to 1,500 BC. Around 600 BC, an editor put the different versions of these events together. The end result is a unique, fascinating, but somewhat incoherent version of what we now consider the Old Testament.

In *Who Wrote the Bible?* Biblical scholar Richard Elliott Friedman notes that putting together various historical documents created a number of conflicts in the final product. Atheists enjoy pointing out these conflicts and inconsistencies. It's easy to do. Biblical scholars, including Friedman, are well aware of the conflicts, particularly in the first five books.

> *It would report events in a particular order, and later it would say that those same events happened in a different order. It would say that there were two of something, and elsewhere it would say that there were fourteen of that same thing. It would say that the Moabites did something, and later it would say that it was the Midianites who did it. It would describe Moses as going to a Tabernacle in a chapter before Moses builds the Tabernacle.*
>
> *People also noticed that the Five Books of Moses included things that Moses could not have known or was not likely to have said. The text, after all, gave an account of Moses' death. It also said that Moses was the humblest man on earth; and normally one would not expect the humblest man on earth to point out that he is the humblest man on earth. (Friedman, 1996).*

Friedman goes on to explain how one of the most striking conflicts is how the editor included two separate stories about how the world was created. The first story has God creating the heavens and earth from nothing; then he creates the fish and animals, and, finally, man and woman. In the second story, God first creates man, then plants, then animals, and, finally, woman.

Many Jews, including *The United Synagogue of Conservative Judaism,* looked at this evidence and concluded that the most popular events from the Bible never occurred. When Biblical scholars claim the Bible is pure fiction, it has enormous implications. If correct, it not only destroys the foundation for Judaism, it also destroys the foundation for Christianity and Islam. It would also suggest that atheists are correct—life is meaningless.

At this point in my quest for God, I became terribly depressed. Amid sleepless nights, I'd toss and turn, wondering about these implications. It appeared my search for God had come to an end. Having rejected the idea of looking inward for answers, religion was all that was left. Without the Bible, I was on the verge of admitting that the atheists were right—there is no God.

Rather than quit, I decided to dig deeper into the work of the Biblical scholars. What I found told me my quest for God was far from over. Instead, it had just taken an exciting new turn.

Chapter 13

REDISCOVERING THE BIBLE

W*hen we want to know something,* we often turn to experts for guidance. The problem with relying on experts is they can often look at the same information and disagree about what it means. Biblical scholars are no different from other experts. Some evaluate the evidence and conclude the major events in the Bible never occurred. Others look at similar evidence and reach the opposite conclusion.

Biblical scholar Richard Elliott Friedman is convinced much of the heart of the Bible did occur. He believes the stories of Abraham, Moses, the Ten Commandments, and the Exodus are valid. He says other scholars who dismissed these events focused on incidental details. By doing so, they missed the big picture surrounding God's message.

Friedman agrees with other Biblical scholars that an editor or redactor put together the histories from two different Jewish tribes. Each tribe had their own traditions and stories surrounding creation, the flood, and other historical events. When the two tribes merged, both versions were kept in spite of inconsistencies.

Why would the editor decide to keep two conflicting stories in the final document? Some suggest he did so to avoid offending the traditions of either group. It's also possible the editor didn't know which story might be correct. Rather than guess, he kept them both.

By combining separate versions of Israel's history into one somewhat coherent document, the editor had to shift events, add transitions, and rearrange stories. This is one reason the end result is so complex and so confusing. But does it work? What does the end product tell us about God and his relationship to us?

Friedman explains how combining sources from different perspectives ends up producing something far richer and far more magnificent than the editor—or any single one of the individual sources by itself—could ever have imagined. By focusing on details, we miss the most important message about the nature of God and his relationship to us.

> *When God creates humans "in his image" ". . . . the Bible pictures humans as participating in the divine in some way that an animal does not. There is something of God in humans . . ."*
>
> *(The editor or redactor) . . . formed a new balance between the personal and the transcendent qualities of the deity. It was a picture of God as both universal and intensely personal. Yahweh was the creator of the cosmos, but also*

"the God of your father." The fusion was artistically dramatic and theologically profound, but it was also filled with a new tension. It was now picturing human beings coming into close personal dialogue with the all-powerful master of the universe. It was a balance that none of the individual authors had intended. But that balance, intended or not, came to be at the heart of Judaism, Christianity, and Islam.

. . . .

Like Jacob at Peni-El (where Jacob battles God), both religions have lived and struggled ever since with a cosmic yet personal deity. That applies to the most sophisticated theologian and to the simplest believer. Ultimate things are at stake, but every human being is told, "The master of the universe is concerned with you." An extraordinary idea. But, again, it was not planned by any of the authors. It was probably not even the redactor's design. It was so embedded in the texts that the redactor could not have helped but produce the new mixture as long as he was at all true to his sources.

. . . .

There was another, even more paradoxical result of the union of the sources. It created a new dynamic between Yahweh's justice and his mercy.

. . . .

Recall that P (one of the original sources) never once uses the word "mercy." It also never uses the words "grace" or "repentance." It never refers to the faithfulness of Yahweh. The priest who wrote it rather emphasized the divine aspect of justice. That is, you get what you deserve. Obedience is rewarded. Transgression is punished. There is no throwing oneself on the mercy of the divine judge.

J and E (two other original sources) are virtually the opposite. They emphasize the divine aspect of mercy.

. . . .

And the redactor combined them. When he did that, he created a new formula, in which justice and mercy stood in a balance in which they had never been before. They were more nearly equal than they had been in any of the source texts. God was both just and merciful, angry and compassionate, strict and forgiving. It became a powerful tension in the God of the Bible. It was a new and exceedingly complex formula. But that was the formula that became a crucial part of Judaism and of Christianity for two and a half millennia.

. . . .

In the combined Biblical text, God is as torn as any loving parent. He makes a covenant with humans, and the contract has terms. When they break the terms, his immediate just response could be anything from termination of the covenant to the arrival of any of the horrible entries on the covenant curse lists in Leviticus 26 and Deuteronomy 28. But his mercy nearly always delays and/or tempers his execution of justice.

. . . .

For those who hold the Bible as sacred, it can mean new possibilities of interpretation, and it can mean a new awe before the great chain of events, persons, and centuries that came together so intricately to produce an incomparable book of teachings.

(Friedman, Richard Elliott, 1996)

There is a reason why the Old Testament became part of the most important, most popular, most widely scrutinized book of all time. It wasn't simply because of the various details associated with Jewish history, or when and how the Earth and man were created. The Bible is the most popular book ever written because it answers the most important questions about God and his relationship to humans.

The Bible is a book about ultimate truth. Whatever its origin, whatever combination of authors and editors contributed to it, the end product turned out to be part of the most magnificent, most impactful book ever written.

The real question with respect to the Bible is not who wrote it, but who other than God could possibly have inspired such a marvelous document?

With these thoughts, I had a new sense of hope. The depression and despair associated with the atheist view began to fade. If God did inspire the Jewish prophets, we should pay careful attention to what they said.

What they said creates a potential problem for many who accept today's conventional knowledge regarding science. Did God really create the world in six days? Were Adam and Eve the first man and woman?

Is there a way to reconcile what the Bible tells us with science?

Let's explore the relationship between science and the Bible.

Chapter 14

SCIENCE AND THE BIBLE

A*theists claim that if you understand science,* you can't believe in God.

As with many of their claims, this one is simply wrong. Brilliant scientists who believed in God and contributed to our current state of scientific knowledge include: Copernicus (Father of Modern Astronomy), Steno (Father of Modern Geology), Pasteur (Founder of Microbiology), Gregor Mendel (Father of Modern Genetics), and Georges Lemaitre (Father of the Big Bang Theory).

Many preeminent scientists see no conflict between science and their strong belief and dedication to God. Science and God are worlds apart. Science deals with the physical world. God deals with the supernatural world.

Conflicts can easily occur when people use the Bible to make statements about the physical world. Theologians are usually not qualified to make claims about the physical world any more than scientists are qualified to make claims about the supernatural.

Over 1,500 years ago, St. Augustine warned Christians not to use Holy Scripture to talk about the physical world. Augustine wrote, "If they find a Christian mistaken in a field which they themselves know well and hear him maintaining his foolish opinions about our books, how are they going to believe those books and matters concerning the resurrection of the dead . . . ?" (Collins, 2006.)

Unfortunately, atheists and believers alike have failed to follow this advice. If the Bible is from God, his intent would have to be to give us instructions about him and our relationship to him. He wouldn't want to give us a science book. What good would it do? It's not as if we could use it to create our own universe.

The most notable conflicts between the Bible and science involve two well-known Biblical stories. Did God create the Earth in 6 days? Or is the Earth 14 billion years old? And, did God create Adam and Eve 5,000 years ago, as a literal translation of the Bible might suggest? Or did our ancestors first appear around 135,000 years ago, as indicated by archeologists and DNA analysis?

In both cases, the discrepancy appears to be more than a rounding error. Or, is it?

Each of these apparent conflicts involves the concept of time. The Bible clearly indicates God is beyond time and space. If God created the universe, he has to be beyond time and space. From God's perspective, the difference between a day and 14 billion years *is* a rounding error.

From God's perspective, our concept of time doesn't represent reality.

This strange idea that our concept of time isn't real is something physicists have discovered only in the past century. Conventional physicists now believe there is another world or dimension beyond our concept of time and space. This revolutionary view of time somehow made it into the Hebrew culture more than 2,500 years before modern scientists came to accept it.

Thanks to Lemaitre, scientists today have confirmed our universe is actually expanding at an accelerating rate. Tracing the expansion back to its beginning, scientists are able to confirm how, going back in time, the universe had been progressively smaller. In the first few seconds of its existence (some 14 billion years ago) it was infinitesimally small. And, prior to its explosive beginning, everything we see or can imagine about our vast universe was nonexistent.

Conventional science tells us every physical thing we can see or imagine on Earth or in the stars came from nothing. Scientists can only speculate about how this could happen. The Bible has an answer—God created it all out of nothing. For many people, believing God created the universe out of nothing is easier than assuming the universe somehow created itself out of nothing.

Scientific evidence for both the time it took to create the universe and for the development of man seems fairly solid. Archaeologists have evidence that humans appeared about 135,000 years ago. DNA analysis for currently living men and women traces our physical characteristics back to a single man and woman.

Were our DNA ancestors the first humans—those referred to in the Bible? Were they the first humans to be blessed with intellect, souls, and knowledge of God? And, did they somehow turn away from God? Science has nothing to say about such questions.

While scientific findings relating to the universe and man are interesting, they don't address life's most important questions: Did God create the universe and man? Did he create them in his image and with an eternal soul? And, do we face consequences based on how we behave?

These are the important questions raised and answered in the Biblical account of creation and the story of our first parents.

Biblical scholars have had difficulty finding evidence of the major events in Jewish history—Abraham, Moses, and the Exodus from Egypt. Did they really happen? Or as some Biblical scholars suggest, are they pure fiction? Science may be able to help resolve some of these questions.

Until recently, Biblical scholars were unable to find evidence to confirm the existence of Abraham, who the Bible tells us was chosen by God to be "the father to many nations." Nor was there any evidence about his sons Isaac and Ishmael, whose descendants (Jews and Arabs) were destined to live in conflict with each other.

Recent discoveries provide evidence of these ancient stories from non-Biblical sources.

Archeologists have found evidence for the journey of a person whose travels closely match those of the Biblical Abraham. DNA analysis also shows today's Jews and Arabs had one father, who lived more than 3,000 years ago.

There is also recent non-Biblical scientific evidence for the Jewish people's relationship to God, the story of Moses, his brother Aaron, the Passover, and the Exodus out of Egypt into Sinai.

In the Biblical story, God sets Moses' brother Aaron and Aaron's sons apart from the rest of the tribe. God chooses them and their descendants to be a special priestly caste known as *Cohanim* (Cohen), with special religious functions. Following the destruction of the Temple in 70 AD, the main role of the Cohen lineage was taken over by the rabbis. However, to this day, the Cohens continue to pass the priesthood on from father to firstborn son and maintain special functions and privileges.

DNA tests from those with specific Cohen characteristics trace the lineage of today's Cohens back 2,100 to 3,250 years. In a fascinating development, some of today's Cohen descendants throughout the world still practice priestly functions. Hence, remnants of the priesthood the Bible says God established some 3,000 years ago remain active today. The evidence indicates these Biblical events did take place. Moreover, they suggest God remains very much involved with the Jewish people to this very day.

As for the Jewish people, no other people in history claimed God spoke to them, telling them they were his chosen people. No other people claim to have received specific directions from God about how to worship him. No other people admit to behaving so badly they were in constant need of God's punishment and repentance. No other people have been as isolated from each other and as dispersed throughout the world. And, after 3,000 years, no other people have managed to be reunited to their homeland, as the God of the Bible had promised.

Science and history cannot tell us if the Bible is the inspired work of God any more than the Bible's stories are designed to help us understand the physical world. The Bible is designed to help us understand ourselves and our relationship to God.

Each of us has to determine the strength of the evidence surrounding God's hand in putting together the Bible. The richness of its stories, complexity of its human experiences, and the advice on how we should live, lead many to conclude the real author of the Hebrew Bible is none other than the Almighty.

What about the rest of the Bible—the New Testament? How does it compare to the Old Testament? Is it valid? Can it help in our search for evidence about God and big questions in life?

For those who find the Old Testament challenging our sense of what's believable, the New Testament takes believability to a whole new level. In one of the few things atheists and believers can agree on—the New Testament is the most incredible story ever told.

Chapter 15

THE NEW TESTAMENT

Throughout the Old Testament, God promises to send a *messiah,* a savior, to Israel and to all the nations of the world. Taylor Marshall's book *The Crucified Rabbi* lists more than 300 Old Testament references to the coming savior. These references, made more than half a century before Jesus' birth, provide extensive details about the *messiah's* birth, actions, crucifixion, death, and resurrection.

Among these predictions: He will be born of a virgin in Bethlehem, be called the Son of God as well as Immanuel (meaning *God is with us*), will speak in parables, enter Jerusalem on a donkey, and will be betrayed for 30 pieces of silver. He is both man and God. Although he is innocent, he will be cruelly tortured and die for the sins of mankind. His hands and feet will be pierced; his body will be struck with a spear. Although he'll die with criminals, he'll have a royal burial; he will come back from the dead and ascend into Heaven.

In so many ways, Jesus fulfills these predictions. He explicitly tells us, "Do not think that I have come to abolish the law or the prophets. I have come not to abolish, but to fulfill."

The unique thing about the description of Jesus' torture and death in the New Testament is how closely his executioners followed the predictions from the Old Testament. The Romans and Jewish leaders had the most to lose by following the predictions for how the *messiah* would die. Yet, that's exactly what they did.

The New Testament is the story of Jesus' life, death, and teachings as told by his closest friends, his apostles. Christians believe Jesus is the promised savior from Heaven and, in some mysterious way, is God made man.

As with the Old Testament, Jesus answers the big questions about God and our role in God's plan. Jesus says God loves us and wants to have a close, loving relationship with his creation. God wants us to love him with all our hearts and souls, place him above all things and show our love by loving others as we love ourselves.

So far, so good.

Then Jesus tells us he wants us to love our enemies. This is difficult. Human nature is to hate our enemies. We want to get even with them. Loving enemies goes against human nature. Jesus often presents us with values contrary to human nature. He tells us God's nature is very different from human nature. God's nature is higher than human nature.

If God created all people in his image, it only makes sense that God values all people, and he wants us to do the same, even to the point of loving our enemies.

Jesus also tells us that, if we live according to God's values, we can expect to have eternal happiness in Heaven. At this point, Jesus seems like a wise, decent, well-meaning person.

Then, Jesus makes it clear how, when we adhere to God's values, it can cause serious divisions within our families, as well as between us and the rest of the world. Jesus also tells us if we fail to follow God's teachings, our souls can endure terrible suffering, burning in an unquenchable fire forever.

Suddenly, Jesus isn't so nice. Although his focus on love and helping others softens his message, his message is the same as in the Old Testament—there are serious consequences for rejecting God. The physics of cosmic justice is real.

In addition to the problem of eternal damnation, Jesus doesn't make it easy to believe in him. Unlike the Old Testament or the Koran, where prophets claimed God and angels talked to them, Jesus not only tells us he came from Heaven, he says he knows God so intimately that he and God are somehow one.

The idea that God, or some part of God, came down from Heaven and took human form defies belief. Anyone claiming to be God is immediately classified as delusional or outright mad. The apostles indicate many who met Jesus, including some in his own family, thought he was mad.

Many are willing to accept Jesus as a wise man and a great teacher, but not as God. The problem is these people don't know Jesus. Jesus claims intimate, firsthand knowledge of God—*Anyone who has seen me has seen the Father. I am in the Father, and the Father is in me.*

Jesus doesn't leave us an easy way out about who he claims to be. He challenges each of us to decide. We can't have it both ways. He is either some delusional madman who we should ignore, or he is so intimately related to God that we'd better listen to him.

Jesus reaffirms the predictions in the Old Testament. He tells his apostles he must suffer severely for the sins of mankind, suffer a terrible death, and then come back from the dead.

On the night before his crucifixion, John's gospel says the high priest asks Jesus directly, "... tell us whether thou art the Christ, the Son of God?" Jesus answers simply, "I am." Jesus' answer is so clear, it causes the high priest to fly off the handle, ripping his clothes and insisting that Jesus must be crucified for the blasphemy of pretending to be God.

No wonder so many considered him mad. It's one level of madness to claim to be God, quite another to claim you will die for the sins of mankind, still another to say you will come back from the dead!

Why did Jesus have to die such a violent death? His closest friends couldn't understand. It wasn't until after his resurrection they finally realized why. There was no way the death of a mere mortal person could somehow redeem mankind from its sins and open the way to Heaven. Jesus, as God made man, was the only one powerful enough that his suffering and death would redeem mankind. His brutal torture and death also were necessary to show how the worst suffering and death anyone could imagine was not the end. Rather, his resurrection would provide proof of life after death.

After his death, only those apostles who first saw him alive believed it. The apostle Thomas wasn't with the others when Jesus first visited them after his resurrection. Even though Thomas knew both Jesus and the other apostles, he steadfastly refused to believe Jesus rose from the dead. He would not believe until he could see the wounds on Jesus' side and hands and touch them with his own hands. The following week he did so. Only then did Thomas kneel before Jesus saying, "My Lord and my God."

Knowing how difficult it could be to believe in the resurrection, Jesus responded, *You believe because you have seen me. Blessed are those who believe without seeing me.*

If Jesus' closest friends had trouble believing it, it's easy to see how people 2,000 years later can also have trouble. The story of Jesus' death and resurrection *is* the most incredible story ever told. Yet, the very scope of such an outlandish story should give us pause to reflect on it.

The question is not simply: *How can we possibly believe it?* If it weren't true, how could his followers believe it? How could more than 500 eyewitnesses believe it? How could millions of people over the past two millennia be so certain it's true that they're willing to devote their lives to spreading the word, and even die for their beliefs?

There can be no doubt Jesus' apostles believed they had seen God. They couldn't wait to tell others about it. The apostles were no longer concerned about dying. They welcomed it. They also insisted on using whatever time remained before their death to spread their good news to the rest of the world.

Tradition has it that, with the possible exception of John, all the apostles were brutally martyred for their beliefs. In all of history, it's difficult to imagine one person dying for a lie, no less so many. The apostles knew Jesus had died. They knew he rose from the dead. Their martyrdom and the martyrdom of countless others over the past two millennia has been an inspiration for others to believe and trust in Jesus. With that trust comes a unique sense of freedom . . . freedom from the fear of death.

The entire validity of the New Testament rests on the story of Jesus' resurrection. The difficulty is, who could possibly believe such an outlandish story? It's a fair question, one people have pondered over for the past 2,000 years.

Did Jesus rise from the dead? Let's begin to examine the evidence.

Chapter 16

DID JESUS RISE FROM THE DEAD?

I*t is the most important question* in Christianity. If the resurrection didn't occur, Christianity is a fraud. If it did occur, the evidence for God is stronger than ever.

Lawyer, author, and historian Ron Tesoriero makes the legal case for Jesus' resurrection. He presents a trial where the jury has to consider if the resurrection occurred. Imagine you are a member of the jury. Listen closely to the evidence before rendering your verdict. Here is Ron's description of the case for Jesus' resurrection.

> *"As Professor John Montgomery, founding member of the World Association of Law Professors, explains, 'Ladies and gentlemen of the jury: The question we have to face is a very simple one. On Easter morning, did Jesus Christ, who was crucified three days earlier, rise from the dead? Now we mustn't make this more complicated than it is.*

The problem is not a matter of knowing what a miracle is or being able to define 'resurrection.' None of us has had contact with such things and is therefore certainly not in a position to provide definitions. But that doesn't mean we are unable to make this case and to make it powerfully.

"While the case for the resurrection is simple, the mechanism, the how, of resurrection is not. But no understanding of how it is possible is necessary to prove that it happened. All that would be required in a court of law to prove a resurrection occurred would be to prove two points: one, that a man died on Friday and two, that on Sunday the same man was alive again. This assumes that we know, with reasonable certainty, the difference between a living man and a dead one. There is nothing to suggest that people two thousand years ago were in any way deficient in this knowledge or that it is the kind of knowledge which relies on modern technology.

"The fact that Jesus was well and truly dead after crucifixion was attested to by primary eyewitnesses. Adding strength to their evidence is the renowned efficiency of the Roman military. In following the orders of their commander, the Procurator Pontius Pilate, the Roman execution squad knew exactly what they were doing. They had crucified hundreds of men before they crucified their most famous prisoner. They made sure that he arrived alive at the point of crucifixion by commandeering a curious onlooker to share the weight of the cross lest he succumb beforehand. To make sure he was dead, whilst still hanging on the cross,

a Roman centurion called Longinus, from the Italian town later named after his lance, pierced the side of the crucified carpenter and ruptured the pericardium of his heart, from which flowed blood and lymph, a confirmation that he was absolutely dead. Only then was permission requested of the Roman authorities for the corpse to be taken from the cross and buried.

"Because the Jewish authorities sought to suppress rumours relating to prophetic statements he had made claiming he would rise from the dead in three days, they took additional precautions to make sure the body remained undisturbed in the tomb. It was sealed with a Roman military seal and guarded overnight by armed soldiers. Breaking a seal enforced by the Roman military carried the penalty of upside-down crucifixion.

"Yet in spite of the best laid plans, the tomb was found empty. The dead man, Jesus, presented himself alive to many eyewitnesses in Jerusalem. He was not, they insisted, an hallucination or a ghost. He was physically present. He ate roasted fish, talked, walked with two of them for more than seven miles and taught them how to interpret Jewish sacred texts. One of the witnesses even put his hand into the wound caused by the soldier's lance. In a space of forty days, more than 5 hundred people witnessed him alive after death by crucifixion. All the crucial firsthand primary evidence for resurrection is positive. If each of those 500 people were to testify for only 6 minutes, including cross-examination, you would have 50 hours of firsthand testimony. Add to

this the testimony of many other eyewitnesses and you would well have the largest and most lopsided trial in history.

"Without the compelling conviction of witnesses, proclamations of a resurrection in Jerusalem could not have been upheld even for a minute in the face of the two powerful parties most hostile to this embarrassing development: the Roman and the Jewish authorities. They had the means, the opportunity and the power to refute the claims, and they never did. All they had to do was produce a body. And yet they never did. Both Jewish and Roman sources and traditions admit an empty tomb. Those sources range from Josephus to a compilation of fifth-century Jewish writings called the 'Toledoth Jeshu.' This positive evidence from a hostile source is the strongest kind of historical evidence. In essence, it means that if a source admits a fact decidedly not in its favor, then that fact is likely to be genuine. For all concerned, those overjoyed by the resurrection and those hostile to it, the empty tomb was an established fact. No shred of evidence has yet been discovered in literary sources, epigraphy, or archaeology that would disprove it.

. . . .

"The New Testament accounts of the resurrection were being circulated among men and women alive at the time of the resurrection. Those people could certainly have confirmed or denied the accuracy of such accounts. The clarion confidence in the chief Apostle Peter's first post-crucifixion public address is diametrically opposed

to his three previous, sniveling denials of being associated with Jesus. 'Men of Israel,' he shouts, 'Jesus the Nazarene was a man commended to you by God. . . . This man you took and had crucified by men outside the Law. You killed him but God raised him to life . . . he is the one whose body did not see corruption. God raised this man Jesus to life and of that we are all witnesses.' Dr Edwin M. Yamauchi, Associate professor of history at Oxford, Ohio, says 'What gives a special authority to the list of witnesses as historical evidence is the reference to most of the five hundred brethren being still alive. St Paul says, in effect, if you do not believe me, you can ask them. Such a statement in an admittedly genuine letter written within 30 years of the event is almost as strong evidence as one could hope to get for something that happened nearly 2,000 years ago.'

"These beliefs were held with such unwavering conviction that, in the centuries after his crucifixion thousands of his followers preferred barbaric torture and death to their disavowal. Those that died rather than renounce what they knew to be true were called martyrs, after the Greek 'martus,' for 'witness' or 'one who bears testimony.' The earliest martyrs were people who claimed to have actually seen Jesus alive after his crucifixion and were literally witnesses to the resurrection. What cannot be emphasized enough is that those who made such claims had absolutely no expectation of any material gain for their outspokenness. Indeed, they proclaimed their 'truth' at the expense of their own safety and in the face of threats of excruciating torture.

Christian martyrs 'were clothed in the skins of wild beasts, and torn to pieces by dogs; they were fastened to crosses, or set up to be burned, so as to serve the purpose of lamps when daylight failed.'

"It seems preposterous to suggest that anyone would be willing to die for a belief founded on a resurrection event they knew to be a lie because they themselves had created that lie. Even Napoleon, a man known to be dismissive of religion, recognized this difference. "I know men and I tell you that Jesus Christ is no mere man. Between Him and every other person in the world there is no possible term of comparison. Alexander, Caesar, Charlemagne, and I have founded empires. But on what did we rest the creation of our genius? Upon force. Jesus Christ founded His empire upon love; and at this hour millions of men would die for him." (Tesoriero, 2013)

We have heard the evidence. Each of us should render our personal verdict. Is the evidence strong enough to confirm that Jesus rose from the dead? If so, it would seem Jesus is who he said he is: God, made man.

Having examined the evidence for the Bible and for Jesus' resurrection, I had a deeper understanding of why so many believe the most incredible story ever told was true. It was now time to look into the evidence atheists had mentioned, and so quickly dismissed, evidence such as the Shroud and Fatima.

If Jesus was God, or somehow intimately related to God, he had to know how difficult it can be to believe in him and his resurrection. Is it possible he intervened in the world and left evidence over the past 2,000 years to help us believe?

The following section presents my analysis of such evidence. As with the legal case for the resurrection, each of us can weigh the evidence and then make the most important decision of our lives—to believe or not believe.

PART IV

Evidence for Jesus as God

In the old days, when there was less education and discussion, perhaps it was possible to get on with a very few simple ideas about God. But it is not so now. Everyone reads, everyone hears things discussed. Consequently, if you do not listen to Theology, that will not mean that you have no ideas about God. It will mean that you have a lot of wrong ones—bad, muddled, out-of-date ideas. For a great many of the ideas about God which are trotted out as novelties today are simply the ones which real Theologians tried centuries ago and rejected.

—C. S. LEWIS

Chapter 17

JERUSALEM, CA. 33 AD

"Late on a Friday nearly two thousand years ago, a wealthy Jewish councillor purchased a large linen sheet. His purpose was to bury the body of Jesus Christ, at that time hanging dead on a cross just outside Jerusalem's walls. Aided by a friend, he unfastened the body from the cross and hurried it to the rock tomb he owned nearby. They worked quickly because the Sabbath, when any labour had to cease, would shortly begin, with the appearance of the first star. They wrapped the body in the sheet, and then rolled a heavy boulder across the tomb's entrance. Thirty-six hours later the first visitors to the tomb found the boulder rolled back and the body gone—except for the sheet that had wrapped it, which had mysteriously been left behind." (Wilson, 2010)

An *ongoing tension existed* between the people of Israel and the Roman army. After conquering the provinces surrounding Jerusalem, Rome became Israel's most recent oppressor. Pontius Pilate, Prefect of the provinces, was primarily responsible for maintaining law and order. To do so, he had the sole authority to order the execution of anyone.

In carrying out their mission, Roman soldiers were particularly cruel and sadistic. Punishment was usually public and severe, designed to discourage anyone from even thinking of breaking the law. A condemned man would first be scourged with the dreaded Roman *flagellum,* a whip with pieces of metal at the end of its tips. The only limit to the number of lashes given was to make sure it didn't kill the victim.

After the scourging, the victim would be forced to carry a *patibulum,* or crossbar, to the place of execution. There the person would be stripped naked and nailed to the bar he had carried. Nails were put through the base of the hand to the back of the wrists. This was done for two reasons—to keep the body from tearing loose from the cross and to activate certain nerves that produce some of the most excruciating pain imaginable.

The crossbar would then be attached to a shaft or tree. In order to avoid causing a quick death from a failure to breathe, the feet were nailed to the shaft. This would enable the victim to lift his body and breathe, thereby extending the period of suffering. After death, the victim's body would be dumped in a public grave alongside other criminals.

This is what was supposed to happen to Jesus' body. But it didn't.

Jewish law decreed there could be no decent burial for criminals. However, two high Jewish priests chose to violate that law. Joseph of Arimathea, a member of the high Jewish Council, asked Pilate for Jesus' body. Joseph, along with Nicodemus, another member of the Jewish Council, were secret followers of Jesus. In keeping with Jewish burial procedures, they lifted the body from the cross, wrapped it with spices of aloe and myrrh, bound it with linen strips, and wrapped it in a linen cloth. The two then took the body to a new tomb Joseph had purchased. They rolled a heavy stone over the entrance to complete a burial fit for a king.

On the first day of the week, when Simon Peter was told the tomb was empty, he entered the tomb and "*. . . . saw the linen clothes lying there, and the face cloth, which had been on Jesus' head, not lying with the linen cloths, but folded up in a place by itself.*" There is no further biblical mention of what became of these burial garments.

Some believe the burial garments have survived to this day. The purported burial cloth—the Shroud of Turin—today is in the city of Turin, Italy. The purported face cloth is in Oviedo, Spain. If these are Jesus' burial garments, they provide the earliest, most direct, non-Biblical evidence surrounding both his death and resurrection.

How could these items have arrived at their current locations? The possible route taken by the Shroud is speculative. Sketchy ancient Syriac manuscripts mention a disciple of Jesus named Addai or Thaddaeus, who came to Edessa prior to 50 AD. Addai was a revered founder of the Assyrian Church in the East. He came with a cloth showing an imprint of Christ. Eastern Church tradition has it that the cloth was instrumental in helping Addai spread Christianity throughout that part of the world.

As with most of the Roman Empire, Edessa was often involved in persecuting Christians. Whenever Edessa's local rulers rejected Christianity, its followers went underground. For almost 500 years, there is little information about "the cloth." Then, in 525 A.D, during a major reconstruction in Edessa, references reappear of an image of Christ, *"not formed by the hands of man."* It was referred to as *The Image of Edessa.*

Prior to the rediscovery of this image, portrayals of Jesus varied. They typically showed Jesus clean shaven and with short hair. Following the rediscovery of the image in the 6th century, most depictions of Jesus show him with long hair, a beard, mustache, and other features found on the Shroud.

In his book *The Turin Shroud,* Mark Niyr describes how the shroud image may have gone from Edessa to Constantinople before appearing in the village of Lirey, France, in the Middle Ages:

"During the year 943, a Byzantine historian reported that the Savior's features were imprinted on a cloth. In that same year (943), the Byzantine emperor determined to bring the cloth from Edessa to Constantinople (modern-day Istanbul, Turkey). In what must have been one of the most unusual military events in history, his army surrounded Edessa (which had been conquered by Moslems) and promised the emir that the city would be spared, that 200 high-ranking Moslem prisoners would be set free, that 12,000 silver crowns would be paid, and that the city of Edessa would be guaranteed perpetual immunity—all in exchange for the "Image of Edessa" cloth. From the year 944 until 1204, the cloth then resided in Constantinople, where it was then called the 'Mandylion.'

"The French 4th crusade besieged Constantinople on occasions between 1203 and 1204. Crusader Robert de Clari wrote about 'seeing in the Church of St. Mary at Blachernae 'the sydoines [shroud] in which our Lord had been wrapped,' adding, 'on every Friday this raised itself upright . . . so that one could see the figure of our Lord on it.' Subsequent events led to a new attack by the crusaders, who ransacked the city. The shroud cloth then disappeared again for nearly 150 years."

Niyr refers to a letter, allegedly written to Pope Innocent III in 1202 ". . . . which complained that the French crusaders had looted the relics from Constantinople—including specifically the linen which wrapped 'our Lord Jesus' after his death."

The next historical mention of the shroud is in the village of Lirey, France, during the 1350s. For most of its history the shroud was owned by the House of Savoy, which turned it over to the Catholic Church in 1983. "The Church takes no official position on the Shroud's authenticity," Pope John Paul II put it in 1998. "The Church entrusts to scientists the task of continuing to investigate."

The other burial cloth described in the Bible is the facecloth, or *sudarium*. This would usually have been placed over Jesus' head prior to taking him down from the cross. While the history of the Shroud prior to the Middle Ages is somewhat speculative, the travels of the *sudarium* are well-documented. The *sudarium* was in Palestine until 614. When Persia attacked Jerusalem, it was moved to Alexandria, Egypt, and then across North Africa. It entered Spain at Cartagena and worked its way up to Oviedo to avoid Muslim invaders. In 1075 the chest containing the facecloth was opened and an inventory taken of all relics. The *sudarium* has remained at Oviedo for more than a thousand years.

Scientific studies of pollen samples on the *sudarium* are consistent with those found in Jerusalem, North Africa, Toledo, and Oviedo. Pollen samples from the Shroud are dominated by a rare pollen found only in the area of Jerusalem, as well as with pollen found in Edessa, Constantinople, France, and Italy.

Are the Shroud and *sudarium* Jesus' burial garments? If so, what can they tell us about Jesus' death and resurrection?

Let's find out.

Chapter 18

JESUS' BURIAL GARMENTS

J*esus' assumed burial garment,* the Shroud of Turin, may have suffered as much rejection as the body it covered. At the time of its earliest documented showings in Lirey, France, the local Bishop D'Arcis vigorously condemned it as a fraud. Both religious and nonreligious commentators have repeatedly called it a hoax. A seemingly fatal rejection occurred in 1988, when three separate laboratories announced that radiocarbon tests dated the Shroud to the 1300s.

Although not nearly as incredible as the story of Jesus' resurrection, the Shroud's authenticity has been continually resurrected. The inventor of radiocarbon testing rejected the validity of the 1988 tests because they ignored the potential for contamination. To be valid, the tested item has to be free from contamination. After 2,000 years of handling and twice surviving fires, the cloth was highly contaminated.

In 2015, a scientific conference at Padua University in Italy produced three peer-reviewed papers based on new, more accurate, modern dating techniques. All three tests dated the Shroud linen to 90 A.D. (plus or minus 200 years), with a confidence interval of 95%. (Niyr, (2020))

Dating with modern technologies also confirmed the unique nature of the cloth. Flurry Lamber, a historical textile expert, had previously identified the unique fabric as a weave she has seen only once before. It was among textiles discovered amid the ruins of the Middle East fortress of Masada on the coast of the Dead Sea. The Masada cloths were dated to within 70 years of Jesus' birth. The unique stitching of the cloth was never found in medieval Europe. The evidence clearly indicates this unique cloth is from the time of Jesus.

As a result of many other unique aspects, the Shroud has become the most scientifically analyzed relic in all of history. Some of the most prestigious scientists and researchers from around the world have volunteered hundreds of thousands of hours analyzing various aspects of the cloth and its image. Over the past two decades, major advances in technology have enabled scientists from different fields to explore the Shroud in details never before available. These discoveries continue to provide possible insights into the validity of the New Testament. *Shroud.com* lists thousands of books and articles offering expert opinions on this amazing relic.

Scientists who have examined the Shroud have reached several conclusions. It once held the body of a well-proportioned man, almost 5' 11" tall, weighing 175 pounds. The man was fully naked when wrapped in the cloth. The marks, wounds, and body characteristics are anatomically correct. The person wrapped in this cloth was definitely a real person.

The victim's back shows evidence of the horrors inflicted from a scourging. The wounds are consistent with the blows of two separate scourgers inflicting a total of about 120 strokes with round metal pieces at the ends of the straps. The whip-like action tore pieces of flesh, leaving depressed centers with elevated edges. These wounds are entirely consistent with the use of a Roman *flagellum*.

Analysis of the cloth also shows marks on the upper back consistent with the victim carrying a heavy crossbeam weighing roughly 100 pounds. Signs of dirt and abrasions on the victim's face and knees reveal the man struggled to walk, falling several times on his face and knees.

Analysis of the blood shows it is type AB, a relatively rare blood type globally, but more prevalent in people in the Middle East. DNA analysis indicates the blood has X and Y chromosomes, indicating the victim was a fully human male.

Bloodstains also indicate the victim was mercilessly tortured. Various parts of the head were badly swollen. Deep puncture wounds around the crown of the head are consistent with marks from a crown of thorns.

Large nails appear to have been driven through the base of the victim's hands and through his wrists. Medical experts confirm how driving nails through palms would not have been able to support the man's weight. The image shows marks on the victim's feet are consistent with being nailed to a supporting structure.

Mark Niyr's description of how the Shroud's victim suffered on the cross is particularly telling:

> *"The abnormally expanded rib cage and enlarged pectoral muscles drawn in toward the collarbone betray strenuous efforts to breathe. In order to breathe from the hanging position, the crucified victim would have to exhale by painfully pushing off the nail spike driven through the feet, raising the shoulders by pulling on the nails from the wrists, and then expanding the ribcage. Blood trails from the Shroud carefully map together these themes. Dual blood trail streams run parallel along the wrists to the elbows at 65 and 55 degrees—evidence that the wrists were nailed to a crossbeam which required this seesaw effort of raising up and down from the wrist spikes (combined with pushing off the foot spike) in order to exhale for each breath of air (and to shift the pain). A wound on the right side of the chest appears to have been caused by a spear.*

The Bible mentions a gush of water flowing out after Jesus' side was pierced with a spear. Analysis of the Shroud confirms the victim was already dead when pierced with the spear.

The traces of the wounds on the body are entirely consistent with the Bible's detailed description of Jesus' tortures.

In all of history, there has been only one person whose death fits this pattern.

In addition to the evidence of what happened to the victim who was wrapped in the cloth, there are intriguing aspects to the image itself.

Microscopic examination shows the image is *extremely* superficial. Each linen thread is about the size of a human hair and is composed of 200 tiny fibers. The yellowish color that makes up the image does not go deeper than *one-hundredth* of the diameter of a single fiber! This is consistent throughout the image. Beyond the image, the fibers are all white.

The image on the cloth appears faint and blurred to the naked eye. Instead of showing these faint outlines, detailed photos of the image show clear negative or opposite details. It is as if the features of the body were to have burst onto the cloth. Today's highly detailed digitized equipment shows incredible three-dimensional aspects of every element of the body that had been enwrapped in the cloth.

Even with current knowledge and advanced computer technology, imaging experts are unable to agree on how such an image could be created today, let alone 2,000 years ago.

Further evidence the Shroud was originally from Jerusalem comes from traces of pollen and flowers found only in Jerusalem. Rare limestone deposits on the feet of the victim were traced to only two tombs in Jerusalem, the Holy Sepulcher and the Garden Tomb (where Jesus is thought to have been buried). Imprints of Pontius Pilate coins were placed on the eyes of the body, providing further evidence of an ancient Jewish burial custom, as well as the timing of the death.

There are far too many fascinating scientific discoveries from the burial cloth to mention here. For those interested, Niyr's marvelous book, *The Turin Shroud,* summarizes and documents in great detail the latest physical scientific evidence from a Jewish perspective.

The other burial cloth described in the Bible is the *sudarium,* or facecloth. Unlike the Shroud, the *sudarium* does not have an image of Jesus' face. As with the Shroud, it has undergone extensive analysis. The analysis shows that the blood and stains on the *sudarium* occurred while the body was still upright on the cross. A second stain was made an hour later, when the body was taken down. A third stain was made forty-five minutes later, when the body was lifted from the ground. It appears the *sudarium* was pinned to the back of the dead man's head. As with the Shroud, spots of blood are consistent with the head being pierced by a crown of thorns.

Additional evidence for the authenticity of both cloths came in the 1980s, when they were brought together. The analysis shows the marks on the *sudarium* are consistent with the location of those marks on the Shroud. The bloodstains on both the Shroud and the *sudarium* are type AB.

Evidence from analysis of the Shroud and *sudarium* is fully consistent with the Bible's description of Jesus' torture and crucifixion. The discarded linens indicate they were no longer needed by the body.

Let's evaluate the evidence and look at the implications.

Chapter 19

EVIDENCE FROM THE BURIAL GARMENTS

I*s there a reasonable explanation* for the burial garments not belonging to Jesus?

If it weren't Jesus, we have to make a number of strange assumptions. We have to assume the Shroud is some elaborate fraud perpetrated by some despicable individuals. Since the body in the Shroud was confirmed to be a human, the fraudsters would have to have acquired a 5' 11", 175-pound male victim and put him through an ordeal remarkably similar to the Biblical account of Jesus' torture and crucifixion. They would have needed extensive knowledge of the details of Jesus' torture and crucifixion. They then would have to have acquired an ancient Roman *flagellum,* and each would have taken turns scourging the victim close to death with about 120 lashes. After the scourging, they would then have tied a crossbeam to the victim's back, forced him to walk some distance, as he fell several times to his hands and knees. The men would then have to have crucified and murdered their victim.

After the victim died, they would have had to thrust a spear through his side, place the *sudarium* on his head, and carefully wrap the body in a very rare and expensive burial cloth designed for someone they honored and respected. Finally, they would have to have put two Pontius Pilate coins over his eyes.

The contradictions involved in cruelly torturing and killing a person, and then giving him an honorable burial would be, to say the least, bizarre.

Those who believe the garments are genuine wonder: What could have happened to the body the garments covered? Why abandon burial cloths, unless they are no longer needed?

Based on the scientific evidence from these relics, each of us can decide. Is it more likely the burial cloths are some gruesome, elaborate hoax? Or do they provide tangible evidence of the Biblical accounts of Jesus' crucifixion and death?

In his account of the Jewish perspective on the Shroud, Niyr concludes his exhaustive survey of the evidence from the Shroud with these words:

> *". . . . if the man on the Shroud is the Biblical Yeshua of Nazareth, then the Shroud would comport to capture that moment of the resurrection which radiated the image onto the Shroud burial cloth. The more you grow convinced that the Shroud is authentic, at that point the Shroud's message*

> *becomes . . . there is life after death! In fact, it becomes tantamount to PHYSICAL EVIDENCE of life after death: left behind and irradiated upon the physical fibers of the Shroud by the power of the resurrection."*

One of the Shroud's Jewish researchers asked his mother if she thought the Shroud was Jesus' burial garment. She replied, "Of course it's real . . . why else would they keep it?"

For those who doubt the Biblical accounts of Jesus' resurrection, no evidence will be strong enough to convince them that it actually occurred. For those who pore through the detailed scientific evidence from 2,000-year-old burial cloths, it is difficult to avoid concluding that the Shroud provides physical evidence that the most incredible story in history—the death and resurrection of Jesus—did occur.

Chapter 20

EXTREMADURA, SPAIN: 1326

In *the summer of 1326,* Gil Cordero was shepherding his cattle near the Guadalupe River in Spain. Noticing one of his cows was missing, he began searching the surrounding area. He soon discovered the cow, lying dead near the river. As Gil drew his knife in an effort to salvage its hide, the cow suddenly came back to life. A figure of a woman bathed in light floated above him.

> *Do not be afraid. I am the Mother of God, the Savior of the human race. Take your cow and return it to the corral with the others, and then go back home. Tell the priests what you have seen. Tell them also that you are sent to them on my behalf. They must come to this place where you are now and dig where the dead cow was; under these stones, you will find an image of mine. When they unearth it, tell them not to take it nor move it away. They must erect a chapel*

for it. In time, a great church, a noble house, and a great nation will grow around this place.

Gil knelt and began to pray. He then went to the town of Caceres to tell the authorities what had happened. No one believed him. He went home and found his wife, neighbors, and certain others crying. Gil's son had suddenly died. As Gil looked over the lifeless body of his son, he recalled how his dead cow had come back to life. He knelt down and, with sincere devotion, begged:

My Lady, you know the message that I am bringing on your behalf, I believe it to be true that you brought this about: that my son is dead because in this way you will show how marvelous you are in bringing him back to life so that this message of yours that I was sent to deliver will be believed quickly. If that is so, my Lady, I beg you to resurrect him. Here and from now on I offer him to you, to be your perpetual servant in the place where you gave me the grace of appearing before me.

The clergy also prayed. The prayers went unanswered. As they carried the boy's body to the cemetery for burial, the boy sat up erect. He asked to be taken to the mountain of Guadalupe, where he could thank the Blessed Virgin for restoring his life.

When they all went to where the cow had been revived, they dug and found a marble box. Inside the box was a statuette of the Blessed Virgin holding a child. There were also relics and documents relating to the origin and history of the statuette.

The documents revealing the roots of what happened in the Extremadura region go back to St. Luke, the skilled writer and researcher of the Gospel bearing his name. Luke was an original Renaissance man—writer, physician, painter, and sculptor. Early Christian traditions indicate he not only interviewed Jesus' Mother Mary but also produced paintings and sculptures of her. Luke died in Greece at the age of seventy-four. Along with his relics, a statuette of Mary and Child were reportedly buried with him.

In 357, the coffin containing Luke's remains was sent to Constantinople, where the Bishop held the statuette of Mary and Child above his head in a great procession. The Emperor of the Eastern Roman Empire gave the statuette to the future Pope Gregory. In 582 a severe epidemic occurred in Rome. Among its victims was Pope Pelagius II. Gregory became Pope and took the statuette in a procession, praying for an end to the epidemic. When the epidemic ended, Pope Gregory attributed it to the work of Mary. Pope Gregory then gave the statuette to the Archbishop of Seville, Spain, in appreciation for his devotion to the church.

The statuette was venerated at the Cathedral of Seville until the year 711, when Muslim invaders threatened the city. Some of the local churchmen quickly took the statuette along with other relics and headed north to a region of Extremadura. They buried the relics in a mountainous area near the Guadalupe River. The statuette would remain buried there for more than 500 years, seemingly lost forever. The Muslim threat had been repelled. Peace returned to Spain. But the relics remained lost to history, lost until that summer day when Gil received a vision that led to their recovery.

As instructed by Mary, the local people built a temporary chapel to house the statuette. The Cordero family remained close to the humble sanctuary now known as the Shrine of Our Lady of Guadalupe. Soon after, King Alfonso XI visited the chapel and ordered a monastery to be built on the site. After being lost to the world for more than 500 years, the statuette has been housed and revered in the Royal Monastery of Santa Maria de Guadalupe since the 14th century.

Did Jesus' Mother actually appear to Gil Cordero, telling him where the relics were hidden? Or did he stumble upon the relics by accident, and make up a crazy story about an apparition, a dead cow, and his dead son?

If so, what purpose would it serve to make up a story no one would believe? And, if he made it up, how would he have been able to convince his wife, friends, neighbors, and priests that such a fantastic story was true?

Were relics hidden for more than half a millennium discovered without divine intervention? Or did everything happen just the way Gil told it, including the reaction of all that it was simply too difficult to believe, until the evidence compelled them to believe?

Gil's little-known, long-forgotten story from the Guadalupe River in Spain provides little in the way of extraordinary evidence for anyone beyond the immediate area. However, more than 200 years later, another reported appearance involving Jesus' mother occurred, one with implications that would reverberate throughout the world for the next 500 years.

Chapter 21

TEPEYAC HILL, MEXICO: DECEMBER 9, 1531

The chill of a cold, wintery wind met Juan Diego as he made his way along a winding, dusty road. Juan lived with his uncle in a region about five miles north of what is today Mexico City. Juan and his uncle were Aztec Indians who had grown up in the area.

For most of their lives, they lived under the control of Aztec priests who believed one of their gods had transformed himself into a ferocious sun god. This god was believed to be in a continual battle with the moon and stars and in need of human blood to restore his strength. Some forty-five years earlier, when Juan was 13 years old, he'd witnessed the sacrifice of thousands of men, women, and children in a pagan ceremony held to appease the sun god.

In 1519 the Aztecs' lives changed dramatically. Spain's Conquistadors had conquered and controlled this area of

Mexico. Franciscan missionaries had built churches in an effort to convert the Aztecs to their Catholic faith.

While the missionaries struggled to convert most Indians, Juan and his uncle were more than amenable. They found the missionaries' description of a God of love far more appealing than the wanton killing associated with the pagan gods. Taking the baptized name of Juan Diego, his wife and uncle were among the first converts to Christianity.

Today, Juan was hurrying to attend a Mass celebrated in honor of Jesus' Mother, Mary. As he passed a deserted area known as Tepeyac Hill, he heard music and the gentle sound of a woman's voice calling his name, "Juanito, Juan Dieguito."

Juan climbed the hill and saw a beautiful young woman whose garments shone like the sun. She identified herself, "I am the perfect and perpetual Virgin Mary, Mother of Jesus, the true God. . . ." The lady told Juan to go to the local Bishop and request a church be built on this site to honor her. In return, she would show her love and compassion and provide protection for the people.

Tradition has it that when Juan asked the lady her name, she responded in his native language, "Tlecuatlecupe." To the Spaniards, this sounded like Guadalupe, the well-known Marian shrine in Spain.

As a peasant, Juan felt uncomfortable about seeing the Bishop. Apparently, the Bishop wasn't anxious to meet Juan, who had to wait for several hours. When Juan refused to leave, the Bishop relented to see him. As with most Marian apparitions, the Bishop quickly dismissed Juan's story.

On his way home, Juan Diego again encountered the apparition. She insisted he go back the next morning with her request. The next morning, after Sunday Mass, Juan Diego again spoke to the Bishop, who told him he would need proof of such an encounter. On his return home, Juan Diego told the lady what the Bishop wanted. The lady told him to return tomorrow, and she would provide proof.

The next day, Juan's uncle became seriously ill. Instead of returning to the apparition, Juan spent the day with his uncle. On the following morning, thinking he was about to die, the uncle sent his nephew for a priest to administer last rites. Juan Diego tried to avoid the apparition by bypassing Tepeyac Hill. It didn't work. She appeared anyway, assuring him his uncle would recover. She asked him to go to the top of Tepeyac Hill, pick the flowers, and bring them to her in his *tilma*.

Juan knew nothing should be growing in this cold and wintery weather. Even so, he climbed the hill and was surprised to discover the most beautiful, fragrant Castilian roses he had ever seen. He picked them, placed them in his *tilma*, and brought them to the lady. She rearranged them and told him to show these to the Bishop.

As on previous visits, Juan Diego had to wait before the Bishop would see him. When the Bishop agreed to see him, Juan Diego opened his *tilma*. As the roses fell to the floor, the Bishop and those with him fell to their knees. On the *tilma* was a beautiful image of the apparition, an image now known as *Our Lady of Guadalupe*. The Bishop immediately recognized the image as Mary, the Mother of Jesus, but he had no way of knowing the full meaning and importance of what had appeared on the peasant's *tilma*.

To the Europeans, the image contained a number of easily recognized Christian symbols that would be found in any European church. To the Aztec Indians, the symbols and colors had very specific meaning. Juan Diego also recognized the image as the apparition. But he, as well as all Aztec Indians, recognized it as much more—a codex or pictograph. It was a religious message full of meaning to all Aztecs. This was a relic that would reshape the history of Mexico for at least the next 500 years.

Since the lady was standing in front of the sun, it meant she was superior to their sun god. Her foot rested on the crescent moon, meaning she would vanquish the moon, their other deity. Her blue-green mantle represented royalty. She was the Queen of Heaven. The stars across her mantle meant she was greater than the stars of the heavens they worshiped. Most incredibly, the arrangement of the stars on Mary's mantle, clear to the Aztec priests, corresponded to their location that very day, December 12, 1531, not as seen from Earth, but as seen from outer space.

The lady wore the Black Aztec maternity belt, which meant she was with child. Over her womb was a quincunx flower. To the Indians, this meant she was a virgin, who bears the true son of God in her womb. Her fingers pointed to the cross on her brooch, the same as the cross on the helmets of the Spanish soldiers. This meant her God was the God of the Spanish missionaries.

While none of these symbols meant anything to the Spaniards, it was a message to the Aztec Indians, particularly to the Aztec priests. *Our Lady of Guadalupe* provided the clearest, most powerful message imaginable: their pagan religions were false; Christianity was the true religion.

In the months and years that followed, there were massive conversions and baptisms. The Indians destroyed images of their former gods and rejected the idea of human sacrifices as barbaric. They abandoned polygamy and opted for Christian marriages.

Missionaries commented how the converts tended to be more devoted to Christianity than many other Christians. Rather than viewing baptism as some vague transformation, the Indians viewed it as a new way of living. They wanted to worship their new, true God by living humbly in peace and love with others. It has been estimated that, within the coming decade, nine million Aztecs became baptized, one of the greatest religious transformations in the world.

While the Aztec Indians clearly understood *Our Lady's* message, there were other important messages embedded in the *tilma* that were not immediately apparent. Over the past half century, with advances in science and instrumentation, awesome details have emerged, details seemingly out of this world.

Chapter 22

THE MYSTERY OF OUR LADY OF GUADALUPE

As *with the Shroud,* the more scientists analyze *Our Lady of Guadalupe,* the more amazing it appears. Although the millions who first saw the image with their naked eyes could appreciate its beauty, they had no idea of the secrets it possessed.

Among the more interesting aspects of the image is how it is not attached to the fabric. Microscopic analysis shows the image is not a painting. There are no brush strokes. Instead, the image appears to be floating above the fabric, vaguely similar to the image on the Shroud.

While the fabric itself is rough and coarse, woven from threads of a cactus plant, the image itself is smooth. Kodak of Mexico compared it to the smoothness of a photograph.

The colors on the image appear to change when viewed from different angles and in different lights. The image of Mary's face from a distance appears dark, much as the complexion of an Aztec Indian; viewing the *tilma* from a closer vantage point, her skin appears to become more white, similar to that of the Europeans.

One of the most fascinating parts of the image are Mary's eyes, which have been extensively studied by numerous ophthalmologists. While *Our Lady* herself is 56 inches from head to foot, the diameter of her irises is a mere 5/16 of an inch. The ophthalmologists confirm that the eyes correspond to all the laws of optics. The precise curvature of the *inside* of her eyes reflects light and images as they would occur with a normal human eye. These laws of optics were not known until three centuries after the image was created. Not only do the eyes conform to the laws of optics, but more recent microscopic analysis of the eyes has identified images of 13 people reflected in both eyes in the exact proportions as would have occurred when viewed by an actual person.

The images in the eyes appear consistent with the scene at the moment the image was first revealed. It is as if Mary produced an image of herself on the *tilma,* with the scene in front of her captured in the eyes of her image.

The material of Juan Diego's cloak, or *tilma,* is made of cactus threads. They usually disintegrate rapidly after 10 to 20 years. And yet, after almost 500 years, there are no signs of deterioration.

In 1785 workmen cleaning the frame accidentally spilled nitric acid on the fabric. While it left a stain, the fabric remained intact. The image endured another trauma on November 14, 1921. Anti-Catholic radicals placed dynamite in a basket of flowers. The explosion crumbled the marble stairs, destroyed metal candlesticks, and bent the huge crucifix on the altar. The image and the *tilma* were unscathed.

On April 27, 2007, the very day Mexico legalized abortion, a mass was held in the Basilica in which *Our Lady* is currently displayed. When the mass ended, an intense white light appeared on the womb of the image of Our Lady. It glowed for a full hour as those present took pictures and videos of the event. The light appears in the shape of a tiny embryo and comes from the *tilma* in the exact location of the woman's womb. Search for "Our Lady miraculous image abortion" to see the pictures.

As with the Shroud and Our Lady of Guadalupe in Spain, we can attempt to construct practical explanations for this phenomenon.

Consider only some of the many skills a forger would have to possess to create such an image. He would have to have figured a way to place the image so that it was floating above the coarse fabric and make it smooth in a way unlike anything ever seen. A forger needed to have not only extensive knowledge of Christianity but extensive knowledge of Aztec religious symbols. He also somehow had to make a fabric with a brief lifespan last for 500 years and seem indestructible.

A forger also had to be brilliant enough to know the precise constellation of stars on the night the image was created as well as what they looked like from outer space. He also would have to have known all the laws of optics that would not be discovered for several hundred years, and somehow figure out how to microscopically imbue the images of 13 people into the two 5/16-inch eyes of the image.

As with the Shroud, there are far too many marvelous features in *Our Lady of Guadalupe* to present in this brief summary. Those interested in further exploring this fascinating phenomenon will never cease to be amazed.

Those unfamiliar with the evidence will respond much as an artist I spoke to outside the Basilica near Mexico City, while visiting *Our Lady of Guadalupe*. I asked him if he had ever seen the image.

"Of course not," he said dismissively. "It's an obvious fake. I wouldn't waste my time even looking at it."

Each of us can draw our own conclusions. We have to decide either to accept that God sent *Our Lady* as a gift from heaven, or that it was created by some forger, who somehow possessed the supernatural powers of God.

Chapter 23

FATIMA, PORTUGAL: OCTOBER 13, 1917

D*ark clouds filled the sky.* The rocky, rolling hills and meadows surrounding *Cova de Iria* were drenched. It had rained all night. And yet, in this obscure piece of the world in the middle of Portugal, throngs of people streamed toward the *Cova,* drawn in anticipation of a miracle.

For the past five months, three small children claimed to have seen visions of a beautiful Lady from Heaven. Ten-year old Lucia, nine-year-old Francisco, and seven-year-old Jacinta claim the vision first appeared to them on the 13th of May. The Lady told them to meet at the same place on the 13th of each month. As word of the visions spread, more people would appear at the *Cova* each month. Other than the children, no one could see or hear the Lady. Still, unusual phenomena were often reported by those who came.

At the July meeting, the children said the Lady promised to tell them who she was and perform a miracle at the meeting in October. Today was the day for the promised miracle.

Those making their way to the *Cova* included the very religious, skeptics, and committed atheists. Earlier in the week, a celebrated journalist for Portugal's leading newspaper had stirred up interest with a clever satire ridiculing the event.

What happened in Fatima is best described by eyewitness accounts among the estimated 70,000 people who fought the elements on that fateful day. John de Marchi's *The True Story of Fatima* provides many firsthand accounts of what happened:

Maria da Capelinha recalls an immediate prelude to the event:

> *"I remember how it was that day, how difficult for the children for a while. There was a priest whom I did not know, and this priest had spent the whole night here. Just before noon, when I began to notice him, he was saying his Breviary. When the children arrived then, dressed as though for their first Communion, this priest asked them directly what time Our Lady would appear."*
> *"At midday, Father,"* Lucia replied.
>
> Looking at his watch, the priest replied, *"Listen, it is midday now. Are you trying to tell us that Our Lady is a liar? Well, child? Well?"*
>
> He was aggressive, this priest, and impatient with the children, and very suspicious. In a few minutes,

he looked at his watch again. "*It is past noon now,*" he said derisively. "*Cannot all you people see that this is just a delusion? That it is nonsense? Go home, everyone, go home!*"

He began to push the three little children with his hands, but Lucia would not go. She was very close to tears, yet full of faith. "*Our Lady said she would come, Father,*" *Lucia said firmly,* "*and I know that she will keep her promise.*"

The rain continued. By local time it was well past one o'clock. But by sun time, it was precisely noon when Lucia looked to the east. "*Jacinta, kneel down. Our Lady is coming. I have seen the lightning.*"

Witnesses say the faces of the children were mirrors of ecstasy, yet what they see and hear is not for others to see and hear, except through the testimony of the children themselves.

"*What do you want of me?*" asked Lucia.

"*I want a chapel built here in my honor. I want you to continue saying the Rosary every day. The war will end soon, and the soldiers will return to their homes.*"

"*Will you tell me your name?*"

"*I am the Lady of the Rosary.*"

Lucia then explained, "*I have many petitions from many people. Will you grant them?*"

"*Some I shall grant, and others I must deny.*"

This Lady of the Rosary, who is God's Mother, is gentle, but she is serious. She has never smiled.

She is asking for penance. She is talking in terms of Heaven and Hell—a blunt and terrifying equation that so many have comfortably forgotten. She speaks as though after 1900 years, a cross still weighs upon the shoulders of her Son: *"People must amend their lives and ask pardon for their sins. They must not offend Our Lord anymore, for He is already too much offended!"*

"And is that all you have to ask?" Lucia inquired.

"There is nothing more."

Then, the Lady of the Rosary took her last leave of her three small friends. She rose slowly toward the east. The children beheld how she turned the palms of her gentle hands to the dark sky over them, and then, as if this were a signal, the rain stopped; the great dark clouds that had obscured the sun and depressed the solemn day were suddenly burst apart; they scattered; they are rent like a bombed rainbow before the eye, and the bold sun hangs unchallenged in its place, a strangely spinning disc of silver.

Lucia, Jacinta, and Francisco were beholding their Lady. From her upturned hands, strange rays of light were rising, as though to assault and make dim the light of the sun itself.

Lucia cried out a single time, *"Look at the sun!"*

At this moment, the drenching rain suddenly stopped, and the sky became clear. Jacinta and Francisco's father described what he saw, *"We looked easily at the sun, which did not blind us. It seemed to flicker on and off, first one way and then another. It shot rays in different directions and painted everything in different colors. . . . At a certain moment, the sun seemed to stop and then began to move and dance until it seemed that it was being detached from the sky and was falling on us. It was a terrible moment."*

Avelino de Almedia, the Editor-in-Chief of O *Seculo,* an anti-Catholic, Masonic daily newspaper of Lisbon, Portugal, described the scene as follows:

> *. . . one could see the immense multitude turn toward the sun, which appeared at its zenith, coming out of the clouds. . . . It resembles a dull silver disc, and it is possible to fix one's eyes on it without the least damage to the eye. It does not burn the eyes. It does not blind them. One might say that an eclipse was taking place. . . . An immense clamor bursts out, and those who are nearer to the crowd hear a shout: "Miracle! Miracle! Prodigy! . . . Prodigy!" . . . Before their dazzled eyes the sun trembled, the sun made unusual and brusque movements, defying all the laws of the cosmos, and according to the typical expression of the peasants, "the sun danced. . . ."*

Witnesses noted how the sun danced for about ten minutes, moving back and forth. At times it appeared to revolve around its axis like a magnificent flywheel, sending out colored flashes of light encompassing every color of the rainbow. It then suddenly appeared to be heading directly for Earth as the crowd screamed in terror, believing it was the end of the world. The frightened people, watching the sun return to its place in the heavens, had not yet noticed that their drenched and mud-soaked clothing had become clean and dry.

On the evening of that same October 13, Father Manuel Pereira da Silva wrote to his friend and colleague, Canon Pereira de Almeida, the following description:

"The sun appeared with its circumference well defined. It came down as if to the height of the clouds and began to whirl giddily upon itself like a captive ball of fire. With some interruptions, this lasted about eight minutes. The atmosphere darkened, and the features of each became yellow. Everyone knelt, even in the mud . . ."

The impressions of Father da Silva are especially interesting because he had been bitterly and outspokenly skeptical of the entire affair. Faith then hit him with such impact that he vowed, perhaps in reparation for his cynicism, never again to indulge his happy taste for wine. Whatever his motive, it was a promise the good father kept.

Actually, this hypothesis of mass hallucination suffers decisive defeat from an incontrovertible fact: the phenomenon was observed not only at *Cova da*

Iria, but by people who were substantial distances away from there, and by no means in receptive spiritual moods.

The Portuguese poet Alfonso Lopes Vieira observed the bright display from a distance of nearly 25 miles. "*On that day of October 13, 1917*," Senhor Vieira recalls, "*without remembering the predictions of the children, I was enchanted by a remarkable spectacle in the sky of a kind I had never seen before. I saw it from this veranda . . .*"

An interesting document has been left by the late Father Inacio Lourenco, a priest from Alburitel, a village about eleven miles from Fatima. We have ourselves taken the trouble to verify his recollections with many of his surviving parishioners, and especially with the schoolteacher, Dona Delfina Lopes, to whom he refers. Here is Father Lourenco's report:

I was only nine years old at this time, and I went to the local village school. At about midday we were surprised by the shouts and cries of some men and women who were passing in the street in front of the school. The teacher, a good, pious woman, though nervous and impressionable, was the first to run into the road, with the children after her. Outside, the people were shouting and weeping and pointing to the sun, ignoring the agitated questions of the schoolmistress. It was the great Miracle, which one could see quite distinctly from the top of the hill where my village was situated—the Miracle of the sun, accompanied by all its extraordinary phenomena.

Firsthand accounts of what happened at Fatima leave us with two choices.

We can accept the evidence of 70,000 eyewitnesses who claim it occurred and believe an all-powerful God suspended the laws of nature. Among the crowd were many—religious and nonreligious—who had strenuously opposed the possibility of a miracle. None denied it occurred. Nor did any deny it occurred at the exact time and the exact place the vision predicted three months earlier.

Alternatively, we can reject the evidence of 70,000 eyewitnesses, as well as others as far as 25 miles away. Some atheists say the miracle of the sun could not have happened because no one can suspend the laws of nature. Therefore, it had to be a mass hallucination.

We should decide. Does the evidence indicate the miracle at Fatima occurred, or is it more likely those testifying to its occurrence suffered a mass hallucination?

If we believe the evidence for the miracle is credible, we should wonder about what it means. Was God just showing off? Or, was there a purpose? What are the implications? What is the meaning of Fatima?

The next two chapters attempt to provide some answers.

Chapter 24

THE CHILDREN OF FATIMA

U*nderstanding Fatima begins* with the children, ten-year-old Lucia, nine-year-old Francisco, and seven-year-old Jacinta. Their lives and experiences provide a glimpse into something few today could even imagine. For those living in today's secular world, the experiences of the children of Fatima can seem almost as incredible as the miracle itself.

All three Fatima children were from devoutly Catholic families. Sunday mass and daily prayers were mandatory. Parents regularly read the Bible to their children and instructed them in their faith. Priests were greatly revered as representatives of God, which is why they are referred to as "Father." As representatives of God, the priests' word was law.

From their earliest years, these children took religion seriously. Although most children did not receive their first holy communion until the age of ten, Lucia convinced a priest she knew her catechism so well she should receive her first communion at the age of six.

The children had an intense love for Jesus and Mary, and often prayed to them. They felt compelled to tell the truth, obey their parents, and avoid swearing. Even prior to the visions, Lucia told of how, each day, the children would say the Rosary, consisting of approximately 60 prayers. She admits how, prior to the visions, they would often rush through the Rosary to have more time to play. After the visions, there would be no more shortcuts.

The children's main chore was taking care of their sheep. After praying and feeding the sheep, they would play in the pasture. It was on one such afternoon, a year *before* their experience with the Lady, they encountered a series of apparitions. Lucia says she was in the pasture with her cousins when she saw an image of a young man, about fourteen or fifteen years old,

> *"whiter than snow and transparent as crystal when the sun shines through it." On reaching them, the image said: "Do not be afraid! I am the Angel of Peace. Pray with me: 'My God, I believe, I adore, I hope, and I love Thee! I ask pardon of Thee for those who do not believe, do not adore, do not hope, and do not love Thee.' Then, rising, he said: 'Pray thus. The Hearts of Jesus and Mary are attentive to the voice of your supplications.'*
>
> *The angel appeared a second time that summer telling them, "Pray! Pray very much! . . . I am the Angel of*

Portugal. Above all, accept and bear with submission the suffering which the Lord will send you."

On a third and final appearance, the angel prostrated on the ground and prayed with the children, gave Lucia holy communion, and gave the blood of Christ to her cousins. Years later, when she was a nun, and asked to recall what transpired, Lucia wrote:

> *The prayers they shared with the angel were indelibly impressed upon their minds. They were like a light which made us understand who God is, how He loves us and desires to be loved, the value of sacrifice, how pleasing it is to Him and how, on account of it, He grants grace of conversion to sinners. It was for this reason that we began, from then on, to offer the Lord all that mortified us, without however seeking out other forms of mortification and penance, except that we remained for hours on end with our foreheads touching the frond, repeating the prayer the Angel had taught us.*

The children told no one of these visions. Lucia would later explain why:

> *The presence of God made itself felt so intimately and so intensely that we did not even venture to speak to one another. . . . The very Apparition itself imposed secrecy. It was so intimate that it was not easy to speak of it at all.*

Their silence concerning the apparitions of 1916 gave the children a year of peace. The following year, the children found out how fortunate they were to have not said anything about the angel. Their year of peace came to an abrupt end.

On May 13, 1917, the children were in the pasture, tending to their sheep when they saw flashes of lightning. They assumed it was about to rain, when as Lucia described,

> *. . . we beheld a Lady all dressed in white. She was more brilliant than the sun. . . . We were so close, just a few feet from her, that we were bathed in the light which surrounded her, or rather, which radiated from her. Then Our Lady spoke to us:*
>
> *"Do not be afraid. I will do you no harm."*
>
> *Lucia asked, "Where are you from?"*
>
> *"I am from Heaven."*
>
> *"What do you want from me?"*
>
> *"I have come to ask you to come here for six months in succession, on the 13th day, at this same hour. Later on, I will tell you who I am and what I want. Afterwards, I will return here yet a seventh time."*
>
> *"Shall I go to Heaven, too?"*
>
> *"Yes, you will."*
>
> *"And Jacinta?"*
>
> *"She will go also."*
>
> *"And Francisco?"*
>
> *"He will go there too, but he must say many Rosaries."*
>
> *Lucia then remembered two of her friends who had recently died.*

"Is Maria Neves in Heaven?"

"Yes, she is."

"And Amelia?"

"She will be in Purgatory until the end of the world," the Lady said.

"Are you willing to offer yourselves to God and bear all the sufferings He wills to send you, as an act of reparation for the sins by which He is offended, and of supplication for the conversion of sinners?"

"Yes, we are willing."

"Then you are going to have much to suffer, but the grace of God will be your comfort." As she pronounced these last words ". . . the grace of God will be your comfort," Our Lady opened her hands for the first time, communicating to us a light so intense that, as it streamed from her hands, its rays penetrated our hearts and the innermost depths of our souls, making us see ourselves in God, Who was that light, more clearly than we see ourselves in the best of mirrors. Then, moved by an interior impulse that was also communicated to us, we fell on our knees, repeating in our hearts: "O most Holy Trinity, I adore Thee! My God, my God, I love Thee in the most Blessed Sacrament!"

After a few moments, Our Lady spoke again:

"Pray the Rosary every day, in order to obtain peace for the world, and the end of the war."

Then she began to rise serenely, going upwards, toward the east, until she disappeared in the immensity of space. Lucia was the only one of the children to talk to the apparition.

The children agreed they wouldn't tell anyone of this apparition. However, Jacinta couldn't contain herself. That night, she told her parents what had transpired. From that point on, the children's lives became a living nightmare.

Chapter 25

THE MEANING OF FATIMA

I*magine the dilemma* confronting these children. They knew the incident was real. They knew it was a sin to lie about it. And yet, their family and others insisted they had made up the story and should confess to lying. Lucia's Mother was particularly adamant, continually insisting she must confess or suffer the consequences.

At the meeting on June 13, about 50 people showed up. When the vision appeared, Lucy asked the Lady to take them to Heaven. She told the children,

> *"Yes. I will take Jacinta and Francisco soon. But you are to stay here some time longer."*

Francisco and Jacinta were overjoyed at the prospect of going to Heaven so soon. Lucia was disappointed she would not be going with her cousins. (Francisco died in the flu epidemic the following year. Jacinta died a year after her brother. Lucia became a nun and died in 2005 at the age of 97.)

Intense questioning by local authorities and priests made the children's lives miserable. The local priest told them their stories were the work of the devil, who had tricked them. Lucia was particularly disturbed by this possibility.

As word of the encounters spread, the number of people joining the children on the 13th of each month increased in size. As the crowds increased, so did the pressure on the children.

The strain was particularly difficult for Lucia. The site of the apparitions was on her family's property, a portion of which was used as a garden for the family's food. Crowds trampled their garden, creating serious financial problems for the family. This caused not only Lucia's mother but also her siblings to berate her for undermining the family's well-being. Frustrated with the belief her child was telling lies, Lucia's mother beat her on at least two occasions.

At the meeting in July, Lucia asked the Lady, "What do you want of me? Who are you? Can you perform a miracle so others will believe you are real?"

The Lady replied,

> *Continue to come here every month. In October I will tell you who I am and what I want, and I will perform a miracle for all to see and believe. I want you to come here on the 13th of next month, to continue to pray the Rosary every day in honor of Our Lady of the Rosary, in order to*

obtain peace for the world and the end of the war, because only she can help you. Sacrifice yourselves for sinners, and say many times, especially whenever you make some sacrifice: 'O my Jesus, it is for love of Thee, for the conversion of sinners, and in reparation for the sins committed against the Immaculate Heart of Mary.'

Lucia recalls,

The Lady opened her hands, as she had in the preceding months, but instead of the glory and beauty of God that her opened hands had shown us before, we now were able to behold a sea of fire. Plunged in this flame were devils and souls that looked like transparent embers; others were black or bronze, and in human form; these were suspended in flames which seemed to come from the forms themselves there to remain, without weight or equilibrium, amid cries of pain and despair which horrified us so that we trembled with fear. The devils could be distinguished from the damned human souls by the terrifying forms of weird and unknown animals in which they were cast."

The children were terrified by what they saw; the apparition looked at them and said so kindly and so sadly, "*You have seen Hell, where the souls of poor sinners go. To save them, God wishes to establish in the world devotion to My Immaculate Heart. If what I say to you is done, many souls will be saved and there will be peace.*"

The harrowing incident of seeing Hell had a chilling effect on the young children. Afterward, seven-year-old Jacinta told them,

Lucia! Francisco! We mustn't stop our prayers to save poor souls! So many go to Hell!

Lucia told how Jacinta was deeply troubled over why people had to go to such a frightful and hideous place as they had seen. She asked,

"Lucia, what sort of sins do they commit?"

"I really don't know, Jacinta. Missing Mass, I guess. Stealing, swearing, cursing. . . ."

"Just for that they go there?"

"Well, I suppose so; it's a sin."

Afterward, Jacinta would tell Lucia,

"I'm thinking of Hell and of the poor sinners who go there," Jacinta said. "Oh, Lucia, how sorry I am for all those souls. The people burning there like coals, I wonder—well, why doesn't Our Lady show Hell to those people who sin? If they could see it, wouldn't they stop? Lucia, why didn't you ask Our Lady to show Hell to them?"

"I didn't think of it."

Jacinta's question from a seven-year-old is the same question we might all ask. Why doesn't God simply show us either Hell or himself and settle things once and for all?

The Gospel (Luke:16) might offer an answer. Jesus tells of a rich man being tormented in Hell for failing to help a beggar. After pleading with Abraham for mercy, the rich man realizes it is too late for redemption. He then pleads with Abraham to warn his five brothers of the torment that awaits them. Abraham's reply:

They have Moses and the Prophets; let them listen to them.

No, Father Abraham, but if someone from the dead goes to them, they will repent.

Abraham replied,

If they do not listen to Moses and the Prophets, they will not be convinced even if someone rises from the dead.

On August 13, the children were unable to meet the Lady. They had been taken captive by the local authorities and questioned incessantly. They were told they would have to admit to lying, or they would be sent to prison. When they refused, they were locked up among prisoners and told they would be killed. They still refused.

While among the prisoners, seven-year-old Jacinta would cry and tell the others, "Neither your parents nor mine have come to see us. They don't bother about us anymore!" The children attempted to overcome their grief by offering their suffering to God, consistent with the advice they were given by both the Angel and the Lady. Once the authorities realized the children would die before changing their story, they returned them to their parents.

As with all other spiritual phenomena, we can each draw our own conclusions about what happened at Fatima. For those who claim the 70,000 eyewitnesses must have been mistaken because the laws of nature cannot be suspended, there are no implications.

For others, there are several things worth recalling. The first is how the miracle occurred at the exact time and place accurately predicted three months earlier. It occurred in modern times, where cameras and photos captured the phenomenon. Following the event, newspaper accounts from around the world documented what had occurred. Photos and stories from skeptics, atheists, and other hostile witnesses corroborated what happened.

Newspaper accounts following the event are readily available on the Internet for those interested in looking into the phenomenon. They clearly show someone or something suspended the laws of science as we know them, in front of 70,000 witnesses.

Evidence from Fatima led me to pay close attention to the children's testimony. It provided significant details about God and what he wanted. These insights centered on the importance of prayer, of talking to God. There was a strong emphasis on how we should pray for others, particularly those who are most in need of our help. The implication is that our prayers are so powerful they can have a very positive effect on the well-being of others.

As in Extremadura, Spain, and Tepeyac Hill, Mexico, the children's testimony indicates Jesus' Mother wants a church built to honor her. Mary says her heart and Jesus' heart are suffering over the behavior of sinners. She wants us to pray for reparations and reform, asking us specifically to say the Rosary every day. She also tells us to sacrifice and deny ourselves certain things. She wants us to offer our sacrifices and sufferings as reparations to those who are in need of our help and in danger of eternal suffering.

In the children's testimony, Mary confirms the existence of Purgatory, a waiting room for Heaven. Our natural inclination is to reject the idea of Hell and be horrified at Mary for frightening young children with such horrors. However, if Hell is real, and every bit as horrific as the children describe, ignoring it won't make it go away.

Another insight from Fatima is how getting to Heaven can be a challenge. As innocent and holy as Francisco appears to be, he was told he would have to say many Rosaries to get to Heaven. It appears only saints go directly to Heaven. As for the rest of us, we'll be very fortunate just to make it to Purgatory, where we'll find out what it takes to make it through the Pearly Gates.

Chapter 26

MARIAN APPARITIONS

For *some reason, Marian apparitions,* as in Extremadura, Mexico City, and Fatima, are the vast majority of all apparitions. A small number of apparitions include Jesus, but an overwhelming number involve Mary.

Of the thousands of reported Marian apparitions, only 17 have been investigated and declared worthy of belief by the Catholic Church. Our Lady of Guadalupe and Fatima have been approved. There was insufficient documentation to approve the event in Extremadura. Among the most famous approved Marian apparitions are those to St. Catherine Labouré and St. Bernadette.

In November 1839, Mary appeared in Paris to a humble Sister of Charity, Catherine Labouré. In the vision, Our Lady told Sister Catherine about the evils of the world and instructed her to have a medal created with specific characteristics of Mary's likeness. It was to be worn by all to commemorate Mary, to safeguard believers against the devil, and to receive special graces. After two years of deliberation, the medal was struck and produced.

Catherine Labouré, whom Our Lady used as an agent, remained unknown to the world during her lifetime. She refused to tell even her fellow nuns. They were unaware of her role as an intermediary in God's work until after her death in 1876. She was canonized Saint Catherine Labouré in 1947, and today her mortal remains lie, still incorrupt for all to see, in the chapel in Paris where the Blessed Virgin first appeared to her. The medal described by St. Catherine has been associated with so many miracles that it is referred to as a "Miraculous Medal."

On February 11, 1858, 14-year-old Bernadette Soubirous saw a vision of Jesus' Mother at a grotto in Lourdes, France. The visions occurred 18 times, ending that summer. Although only Bernadette could see or hear the vision, the number of spectators grew until, at one point, there were 8,000. The vision finally identified herself to Bernadette as "Que soy era Immaculada Concepciou," (I am the Immaculate Conception). Unknown to Bernadette, this was a theological expression adopted by Pope Pius IX four years earlier as Catholic dogma. As in other Marian visions, Mary asked for a chapel to be built on the site.

Bernadette became Sister Marie Bernard and spent the rest of her life serving the sick. She died at age 35 and was declared Saint Bernadette in 1933. The miracles associated with Lourdes are well-known.

Marian apparitions provide fascinating insights to those who receive them. Detailed accounts of such apparitions are available in books and over the Internet for those interested in learning more about them.

While Mary appears to have been busy making appearances, she may not be the only member of the Holy Family who has visited Earth.

Chapter 27

SANTA FE, NEW MEXICO: 1852

Sister Magdalen Hyden was among six brave young volunteers from the *Sisters of Loretto at the Foot of the Cross* in Kentucky. They answered the call for teachers to help the poor in the rough and dangerous territory of Santa Fe, New Mexico. Warned of the dangers involved, Sister Magdalen was one of the four who survived the three-month journey to Santa Fe.

Due to her kindness and generosity, the Irish nun was chosen to be the new superior. She spent the next twenty-nine years establishing schools from Colorado to Texas. By 1873, the sisters had saved $30,000 toward their goal of building a chapel. The project was plagued with numerous problems. It was nearly completed in 1878, when its builders pointed to a major problem. The choir loft was so high, there was no way to build a staircase to it without having the stairs intrude well into the rest of the structure.

Builders told the sisters the only solution was to tear out the loft and lower it. They refused. Instead, they started a Novena to St. Joseph, the patron saint of carpenters. A Novena consists of nine days devoted to prayer with the hope of receiving special graces.

It was on the last day of the Novena that a gray-haired man, leading a donkey and bearing a tool chest, arrived at the convent gate. Upon asking to see the superior, he told Mother Magdalen he was a carpenter and would like to build the much-needed staircase. She was so eager to agree, she failed to ask his name or where he was from. Sisters at that time say the only tools he used were a hammer, a saw, a T-square and a few other hand tools he kept in a small chest. They also recall seeing wood soaking in tubs of water as he worked on the staircase.

Upon completing the project, the gray-haired man disappeared, leaving an architectural marvel.

> *There are no central column or support beams, and it appears that all the weight is self-supported at the base. The craftsman did not use nails or glue; he only used wooden pegs to secure the steps. Additionally, there were no railings. The legend says that some of the nuns were so afraid to descend the 22-foot drop that they would crawl down on their hands and knees. There are only 33 steps, however, the staircase wraps around 360 degrees twice.—historicmysteries.com*

Builders and architects are amazed the structure doesn't collapse the first time someone steps on it. And yet, it has held fifteen people at one time.

Historian Mary Cook researched the stairs and found a worldly explanation for the construction. She found an entry in 1881 in the nuns' daybook for "wood" to a man named Rochas. According to Cook, Francois-Jean Rochas was a member of a French secret society of highly skilled craftsmen and artisans called the Compagnons, which has existed since the Middle Ages. Cook says that Rochas came to the U.S. specifically to build the Santa Fe staircase.

Did Saint Joseph respond to the sisters' Novena, directly or indirectly? Or, was it just a coincidence that a stranger, with unworldly skills and exotic materials, just happened to appear at the right time, in the right place, to solve a seemingly unsolvable problem?

Each of us can weigh the evidence.

Chapter 28

STIGMATA

You mean go all the way to some unknown godforsaken place in the middle of Bolivia to see if a crazy woman's hands bleed on cue? These things can't happen and don't happen. It's all mind over matter, or a kind of self-hypnosis or psychosomatic hysteria or a couple of religious fanatics needing attention and getting up to tricks. Been there, done that. Don't expect me to believe because you believe.

—MIKE WILLESEE (TESORIERO, 2007)

Mike Willesee had seen it all. A celebrated journalist in Australia, he'd made a career of debunking claims of supernatural events. He wasn't about to listen to Ron Tesoriero's strange request. Ron had been to Bolivia to investigate a bleeding statue of Jesus and a lady supposedly bearing the stigmata—the wounds of Jesus.

Ron Tesoriero wasn't discouraged by his neighbor's objection. He persisted, "You've got a world-class reputation for integrity. You always do a thorough investigation and present an honest evaluation, and so. . . ." After pausing for effect, Ron continued, "Prove me wrong!"

Ron and Mike were both fallen-away Catholics. Mike was still hostile to Catholicism due to events in his past. Ron, a partner in his successful law firm in Sydney, Australia, had long since left his religion behind. Then suddenly in 1987, a series of seemingly miraculous events in his personal life led him to reconsider how God could be intervening in our world.

By 1993, Ron had not only returned to practicing his faith, he'd bought a camera and began traveling the world filming supernatural experiences attributed to the hand of God. In 1995, he received a call from a friend, Professor Ricardo Castanõn, about a strange case of a statue of Jesus that was weeping blood and tears.

Ron quickly grabbed his equipment and was off to Cochabamba, Bolivia, to see what this was all about. What he saw was a small, plaster statue of Jesus' bust. Ron described it as "gruesome and tragic, unnervingly lifelike . . . seemingly etched in sorrow with its own blood." He watched and recorded the formation of what appeared to be tears and blood as they flowed from the statue.

Ron and his friend Professor Ricardo took the statue to a local hospital, where a CAT scan showed the statue was hollow. They then sent samples of the tears and blood to several laboratories for testing. Those analyzing the samples were not told where they came from until after the results were in.

The tests revealed the samples contained both human skin and blood. The blood had all the attributes of fresh human blood, with one mysterious difference. Scientists analyzing the blood were puzzled at not being able to detect a genetic profile. After being told where the blood came from, a forensic pathologist from San Francisco said, ". . . if it was a deception, only blood would be present. Not skin cells as well. It is not within the realm of our understanding. For myself, on a very personal level, I would prefer this to be a trick." (Tesoriero, *2007*)

Professor Ricardo then told Ron Tesoriero of a nearby lady who had been receiving messages in the form of dictation from Jesus. "Katya," or Catalina Rivas, was a grandmother, a simple person who speaks only Spanish and had not graduated from high school.

At age 48, she began writing hundreds of pages of Biblical and theological commentary in Spanish as well as several other languages. She claims the words come directly from Jesus, who wants her to write them down. She says she often doesn't understand the meaning, particularly when it's in another language.

The Bishop of Cochabamba has given his imprimatur to her writings, saying they are without theological error.

Ron says Katya wrote for an hour nonstop as he filmed her. One commentary stood out in reference to the statue:

> *I have taken as a symbol, a piece of wood, a cross. I have carried it with great love for the good of all. I have suffered real affliction so that everybody could be joyful with me. But today, how many believe in him who truly loved you and loves you? Contemplate me in the image of the Christ who cries and bleeds.*

Jesus also addresses the subject of prayer in Katya's writings:

> *There has never been anyone who has placed his faith in Me and has been abandoned. He who prays to me in faith, obtains all he has asked for. When the graces asked for are spiritual and useful to the soul, be sure that you will receive them.*

One of Katya's writings has Jesus explaining what he meant about "blessed are the poor in spirit":

> *We shall talk about the "poor in spirit." I said, "Blessed are the poor in spirit." Actually, the meaning of these words in Hebrew is "blessed are those who lack arrogance." The significance of this phrase has been misinterpreted. These*

words are meant to imply not that I praise solely those unattached to earthly goods but instead, to exalt the humble.

Ron was impressed with everything he encountered in Bolivia. In 1996 he received news from Bolivia that Katya had experienced the stigmata—the wounds of Christ. It was at this point he felt the need to call on his celebrity neighbor. Ron wanted a thorough professional investigation of Katya's stigmata as well as the bleeding statue. He knew Mike was the one either to verify or refute these purported signs from God.

In 1996, Mike and Ron arrived in Cochabamba. Mike was certain he would prove statues don't cry and bleed, and a person can't spontaneously bleed and receive the wounds of Christ.

Mike's first impression of Katya was positive. He found Katya to be humble and holy. He could see there was no money or other ulterior motive involved. However, there was also no stigmata.

Next, they went to see the bleeding, tearing statue. They watched and filmed the statue. Neither tears nor blood oozed out.

To Ron's dismay, their journey failed to produce anything of consequence. Or so it seemed.

Chapter 29

SIGNS FROM GOD

Although their trip to Bolivia appeared to be a failure, some of the films taken there caught the attention of a Fox television executive, who was intrigued by what saw. He called Mike Willesee and offered to hire him to produce a two-hour televised documentary called "*Signs from God.*"

Since people given the stigmata often suffer on Fridays or during Easter, Ron, Mike, and a group of film technicians again left for Bolivia the week before Easter in 1999 to film the documentary.

The day before Good Friday, with all the cameras rolling, Katya said she received dictation from Jesus, "There will be no stigmata tomorrow. He says it will be in His time and you must learn to trust Him." Disappointed, they packed up their equipment and left for the day.

The next day, again with the cameras rolling, Katya said she had another message. "It will occur the day after Corpus Christi, and you will be able to come with witnesses and film what happens. I will have the wounds, which signify the Passion of Christ, on my head, my feet, and my hands. It will start around noon and end just after 3 p.m. And it will heal by the next day."

They all flew back to Australia with what Ron described as, "thousands of very expensive miles," only to return two months later on June 9 for the predicted stigmata.

This time they were not disappointed. They filmed the event, capturing each detail, testing the blood and witnessing the agony Katya experienced, along with her full recovery. The program "*Signs from God*" was shown to massive American and Canadian audiences in July 1999.

At the end of her ordeal, Katya said Jesus spoke to her, saying,

> *I have been preparing you for this day because I needed to reach the world one more time to show the world my suffering through a person like you. Thank you.*

But Jesus wasn't finished with the miracles of his blood.

Three months after Katya's stigmata, Professor Ricardo again called Ron with an urgent request, "Join me in Argentina. Katya spoke to me. She has a message from Jesus for me: '*Tell Ricardo I want him to take charge of this case. Through it I want to bring dignity back to my altars.*'"

Professor Ricardo greeted Ron upon his arrival in Buenos Aires. The Archbishop himself, Jorje Bergoglio (later to become Pope Francis), had asked Ricardo to conduct an examination of a mysterious development—a discarded host that appeared to be bleeding.

On August 18, 1996, Padre Alejandro had just finished celebrating the 7 p.m. Mass when a lady told him she had seen a communion host abandoned in a candle holder. "I picked up the host and took it to the altar with the intention of consuming it. I saw the host was very dirty, and so I asked a minister of communion to put it in a bowl of water and to put it in the tabernacle."

Several days later, the minister looked in the tabernacle and found the eucharist was becoming red. The bloodlike substance continued to grow over the following weeks. The remnants of the host and the bloody red substance were still present three years later, when Ricardo was asked to conduct an examination.

The material was sent to several laboratories for analysis. None were told where the material came from. The tests showed the specimen was heart-muscle tissue and DNA. Although the blood appeared to be fresh, the analysis was unable to detect a genetic profile.

One analyst said, "It looks to be of human origin. . . . The heart muscle is inflamed. There has been recent injury, like those that I see in cases where someone has been beaten severely around the chest." The white blood cells ". . . . indicate injury and inflammation. White cells can only exist if they are fed by a living body. This sample was alive the moment it was collected."

There are other documented cases of bleeding eucharists. On December 8, 1991 in Betania, Venezuela, Father

Otty was saying mass. After the consecration, he broke a large host into four pieces. He consumed one piece and noticed one of the remaining ones began turning red. The local Bishop set up a commission to study the miracle. The bloody eucharist is available to see all the days of the year at any hour in a chapel devoted to Perpetual Adoration. For Catholics, the eucharist is the body and blood of Jesus, which should be revered and adored.

One of the miracles attributed to the event in Betania occurred on November 12, 1998. Daniel Sanford, a young American from New Jersey, was among the pilgrims who came to see the bleeding Host. He tells how, "With great astonishment I saw that the Host was as if in flames, and there was a Pulsating Heart that was bleeding in Its center. I saw this for about 30 seconds or so, and then the Host returned to normal. I was able to film a part of this miracle with my video camera. . . ." (Search "pulsating heart eucharist" to see Daniel's amazing video).

Still another report of a bloody eucharist occurred in Tixtla, Mexico, on October 26, 2006, at the Parish of St. Martin of Tours. One of the ladies distributing communion burst into tears when the host she was about to give began to turn red. The local Bishop commissioned a scientific investigation. It ran from October 2009 to October 2012. The results were presented in 2013. They showed the material was fresh, human blood type AB.

Still another example of a communion host turning into blood and flesh soon followed. The event occurred on

October 12, 2008 at the Church of St. Anthony in Sokółka, Poland. Once again, a commission was set up to investigate. The findings were similar to the others.

The oldest and most extraordinary documented evidence of a bleeding eucharist dates back to the 8th century. A priest of the order of St. Basil held up the altar bread and said the words to transform the host into the body and blood of Christ. In front of the entire congregation, the host began to bleed. The priest, who had doubts about transforming bread into the body and blood of Jesus, immediately proclaimed it was a miracle from God to overcome his doubts.

The host and globs of blood were collected and placed into a container. In 1971, modern scientific tests were conducted on these remnants, which were then more than 1,200 years old. An independent analysis confirmed the presence of flesh from a human heart and blood, type AB. After more than 1,200 years, the blood had all the characteristics of fresh human blood.

In her book *Eucharistic Miracles,* published in 1986, Joan Carol Cruz writes about 36 well-documented examples of bleeding hosts. Catholic teaching maintains the substance of the eucharist becomes the real body and blood of Jesus. However, the properties of the bread and wine remain the same. Hence, if you test a consecrated host, the expectation is that it will have the properties of bread and water, not flesh and blood.

Bloody eucharists that have been scientifically analyzed are the exception. Scientific tests reveal something that defies science. Many believe bleeding eucharists are a message from Jesus.

Once again, we have a decision to make. Did these things actually occur? If so, what is the message?

Chapter 30

THE BLOOD OF CHRIST

A *statue bleeds human blood.* Blood from a host collected more than 1,200 years ago remains fresh. Other hosts turn into fresh human blood. A 45-year-old peasant grandmother miraculously suffers the wounds of Christ at the precise time and place predicted and completely recovers the next day.

What could it mean?

We are left with two possibilities. First, these events represent an elaborate hoax. Bishops, priests, pious volunteers, church attendees, scientific experts, and lab technicians have all conspired to deceive people. But, for what purpose? To lie about such things would be a serious sin. Those involved would see it as condemning their souls to Hell. And, why even try to convince people of something so unbelievable that, even if it were true, few would be expected to believe it?

The alternative explanation is that it really happened. The events are real. They are from God.

But why the emphasis on blood?

Blood played a major role in the Old Testament. The Book of Exodus tells how Moses received instructions from God concerning both the priesthood and the ceremonies God wanted the Israelites to perform. The Israelites were to build a tabernacle, which was to be the earthly dwelling place of God. The tabernacle was to have an inner space—the Holy of Holies—where God was to dwell. A purple linen curtain separated the most holy place from the rest of the Tent of Meetings. Outside the Holy of Holies was an altar where the blood from unblemished sacrificial animals was to be spread in accordance with instructions handed down from God. The blood of the animals was sprinkled on the altar to atone for the sins of the Israelites.

The Bible explains how, at the time of Jesus' death, the veil in the Temple was torn in two from top to bottom. Apparently, the veil separating God from man was no longer necessary.

Christians believe Jesus is the sacrificial lamb sent from God to atone for the sins of mankind. From his birth, he was wrapped in swaddling clothes. These were the clothes used by the Levitical priests, the shepherds mentioned in the Bible. It was their job to wrap the unblemished animals in swaddling clothes to protect them from blemishes prior to the ceremony. At the time of Jesus' birth, an angel told the shepherds in a nearby field: *You shall find the infant wrapped in swaddling clothes, and laid in a manger.* The manger was the trough where the cattle and sheep would feed. The message to the Levitical priests was clear. Their savior had arrived. A savior, the lamb of God, destined to be sacrificed.

The Old Testament of the Bible provides numerous details about how the savior would die a horrific death. From the standpoint of Christianity, Jesus' horrific suffering, crucifixion, and death on the cross were pre-ordained.

At one point in his ministry, Jesus attempts to prepare his followers for who he is. A year before the Passover of his last supper, Jesus was teaching in the synagogue in Capernaum when he told his followers,

I am the bread of life. . . . I have come down from Heaven, not to do my own will, but the will of him who sent me. And this is the will of him who sent me, that I should lose nothing of all that he has given me, but raise it up on the last day. This is indeed the will of my Father, that all who see the Son and believe in him may have eternal life; and I will raise them up on the last day.

Those surrounding him began to complain, saying, "How can he say, 'I have come down from Heaven'?"

At that point, Jesus became even more specific about who he is:

> *I am the living bread come down from Heaven. Whoever eats this bread will live forever; and the bread that I will give for the life of the world is my flesh.*

Most of Jesus' followers were stunned. They argued among themselves, "How can this man give us his flesh to eat?"

Jesus then doubles down and becomes even more specific.

> *Very truly, I tell you, unless you eat the flesh of the Son of Man and drink his blood, you have no life in you. Those who eat my flesh and drink my blood have eternal life, and I will raise them up on the last day; for my flesh is true food and my blood is true drink. Those who eat my flesh and drink my blood abide in me, and I in them.*

Jesus followers were shocked. They complained, "This teaching is difficult; who can accept it?"

With that, many of Jesus' disciples, unwilling to accept what he said about his flesh being true food and his blood true drink, left him. Only the twelve remained.

A year later, during Jesus' eulogy at the last supper, he reaffirms the importance of his body and blood when he passed the bread and cup to them, saying,

> *Take, eat, this is my body; drink all of this, for this is my blood; do this in remembrance of me.*

Why would Jesus deliberately lose many of his followers by stressing the importance of eating his flesh and drinking his blood? And, why would he repeat the same thing a year later during his personal eulogy, the night before his crucifixion and death? And, why would eucharists, which represent the body and blood of Christ, turn into flesh and blood?

These are among the many questions and mysteries that surround Jesus.

Chapter 31

SUMMARIZING AND EXPANDING THE EVIDENCE FOR GOD

At the beginning of my search for God, one of my goals was to accept the evidence wherever it might lead. It led me to the Judeo-Christian roots of the Bible.

More than half a century before Jesus' birth, Old Testament Jewish prophets provided detailed descriptions of a promised savior. When Jesus fulfills these predictions, it provides evidence that the Jewish prophets were relaying messages from a Supreme Being, one with a grand plan for the ages. It also points to Jesus as someone unique from anyone who ever lived.

Jesus tells us he came from Heaven, was so closely related to God that he and God were one, would have to die for the sins of mankind, and would rise from the dead. In all of history, no sane person made such extraordinary claims. Jesus is either a madman or he is who he says he is.

In the 2,000 years after his death, additional evidence points to Jesus as the promised savior. Burial garments, Our Lady of Guadalupe, Fatima, blood miracles, and other phenomena indicate God remains active in our world and cares very much about his people.

For me, any one piece of this evidence might not have been convincing, but all of it combined presents a clear path to God being real.

This evidence may not appeal to everyone. Those with different beliefs and traditions will likely reject some, or even all, of the evidence as inconclusive. This raises important questions.

Did my search reveal the truth about God? Would others, with different talents, beliefs, and traditions find different evidence and different conclusions in seeking God?

Although God led me on a specific path to find him, there may be many paths to finding him. There may be as many different paths to God as there are people. Just as our beliefs and talents differ, our individual searches for God may differ.

Disagreements over details surrounding God and his nature are inevitable. God's nature is so far above our human nature that we will never fully comprehend it. Whatever we choose to call him—El, Yahweh, Lord, Father, Yeshua, Jesus, Allah—we are calling on the same God. We are calling on the one who created the heavens and the earth and all of mankind. However well or imperfectly we might understand him, he remains our God.

If you're searching for God, it may help to realize he knows us better than we know ourselves. He knows of our traditions, our beliefs. He wants us to find him and share in his joy and peace. So long as we seek him with an open mind, with honesty and humility, we have God's promise—*all who seek shall find, and to all who knock, the door will be open.*

If you have already found God, that's wonderful! Each person who discovers God has insights and evidence that can help others find him. If you have found God and are able to share your story, consider going to the website—myevidenceforGod.com. Tell your story. Let others know about your journey and your evidence for God. Stories and evidence from those who have found God can help inspire others to undertake the most important journey of a lifetime—the journey to discover evidence for God.

PART V

Lessons from My Search for God

Thou hast made us for thyself, O Lord, and our heart is restless until it finds its rest in thee.

—ST. AUGUSTINE

Chapter 32

REFLECTIONS ON MY LIFE—CHANGING SEARCH FOR GOD

When I began my search for God, I had no idea where it would end, or of the challenges and barriers I would face. Looking back, I realize how the unpleasant pitfalls, blind alleys, and harrowing times served to enrich and deepen my faith. As the project drew toward an end, it became apparent God had made it relatively easy.

God left evidence all over the place—much as a child who plays in the mud and then wanders into his home. God left his footprints everywhere—in historical events, stories long forgotten, and places few might think to look. While any search for God never ends, I felt a strong urge to draw it to a conclusion, present my findings, and move on.

The good news is—atheists are wrong. God not only exists, he loves each and every one of us. God loves Jews, Christians, Muslims, Bhuddists, Hindus, pagans, and yes, he even loves *atheists.* If he didn't, he wouldn't have created us.

Searching for God has been a labor of love. It has rewarded me with insights, inspiration, and a peace far beyond anything I might have imagined. The greatest benefit has come from a sense of moving closer to God. While this move is still in a beginning stage, it seems slight movements in God's direction can produce amazing benefits. As the ultimate expression of love and peace, any move closer to God brings me closer to a wonderful feeling of serenity.

Experiencing such peace has made me more attentive to what happens when we move away from God. I've become more sensitive to seeing those around me caught up in the turmoil of this world. It reminds me of the many times I would become anxious, distressed, and even fearful over various crises. I now realize, my real problem was that I was moving away from God.

By moving away, I had let myself become overly immersed in the chaos of this world. Although the details of my problems were constantly changing, the anxiety associated with them seemed to be a permanent condition.

Jesus tells us, *My Kingdom is not of this world.* His world is different, free from our worldly concerns. The more I was able to bring God into my life, the more the seemingly unending troubles of this life felt less important.

Fear is destructive. Jesus tells us, *Do not be afraid.* Biblical scholars tell us the Bible tells us not to be afraid 365 times, one time for every day of the year. The more fearful we are, the further removed we are from God.

During my search for God, atheists have been particularly insightful. Many atheists have not only moved away from God, they appear to despise, even hate, both God and those who believe in him. Hatred is a destructive consequence of having moved far from God's love.

Without God, our lives are without meaning, purpose, and direction. Atheists who confront the implications of such a life tell us of their neuroses, their fear of death. Without God, it's easy to devolve into a state of emotional instability searching for therapists, medication, drugs, alcohol, the metaverse. Their real search is for anything to use as an antidote to a meaningless world—an existence without God.

Jesus guides us to a more fruitful life. He provides us with a clue to what he offers when he says, *Peace I leave with you; my peace I give you. I do not give to you as the world gives.*

The more I attempt to make God a part of my life, the more my life is filled with purpose and direction. Hatred is untenable. I become more confident, focused, optimistic, hopeful.

Getting to know God involves building a relationship. Good relationships take time and commitment. Making time for God means less time for other things. One thing Jesus tells us to do is to put God first in our lives.

> *Love the Lord your God with all your heart and with all your soul and with all your mind. This is the first and greatest commandment. And the second is like it: Love your neighbor as yourself.*

Putting God first, above family and friends, is difficult. It's something I still have to work on. However, simply elevating God's importance has helped me deal with potential crises, which often fade into the background and dissipate. Problems which once seemed overbearing become far less significant when compared to Jesus' suffering on the cross and the pressing needs of so many others.

One of the most important changes in my life has been a better understanding of those terrible moments we all must face—tragedy, morbidity, our own mortality, and the deaths of our loved ones.

Almost a century before Jesus' birth, the psalmists and prophets provided numerous details about how the coming *messiah* would have to die a horrible death to save mankind. God revealed not only his plan to save the world, he also revealed the gory details of how his Son would suffer. Only after the worst suffering and death imaginable would the resurrection provide two crucial lessons for mankind—*suffering can be helpful* and *there is life after death.*

Like so many, I wanted to avoid dealing with or even thinking about suffering. Jesus' suffering tells us God's world is very different from ours. As humans, we do almost anything to avoid suffering. We bury it deep in the recesses of our mind, as if to avoid talking and thinking about our inevitable death or the death of loved ones can prevent it.

In God's seemingly upside-down world, the worst suffering imaginable is fruitful. It brings about salvation. That doesn't mean it isn't painful. It is. We can't avoid pain and suffering, but we can react to it in one of two ways. We can sink into a deep depression, view our loss as the worst thing in the world, and dwell on it uncontrollably, while continually asking . . . *Why me? What have I done to deserve this?*

Many of the saints, those closest to God, have an alternative response. They focus on Jesus' suffering on the cross. Since there is no suffering in Heaven, on earth it is viewed as the only opportunity to relate to Jesus. Many saints welcome suffering. They see it as an ideal chance to exercise the empathy they feel for what Jesus endured for us.

Some saints believe glory in Heaven can be directly proportional to how much we endure here on Earth. They actually pray for suffering to bring them closer to Jesus. In modern times, the thought of praying for pain and anguish seems ridiculous. In God's world, it isn't.

In his book *When You Suffer,* Jeff Cavins provides us with the alternative approach. Cavins suggests we begin preparing ourselves for those terribly painful moments we all must eventually encounter. We can start by offering up to God the more trivial things that go wrong for us. When we become upset, annoyed, or mad at someone, we can offer our suffering to Jesus as a small measure of our gratefulness for his suffering for us. The more difficult the day, the greater our disappointment, the greater the value of our offering.

This perspective on suffering can keep us from becoming preoccupied with ourselves and our sorrows. As we grow closer to God, we begin to appreciate how even our worst tragedies can have a happy ending, one where we and our loved ones are reunited in Heaven forever.

Moving closer to God dramatically changes my response to so many things. It allows me to deal, not only with adversity, but to prepare for those terrible moments we are all destined to encounter. With God as my rock, the words from Robert Lowry's famous song take on deeper meaning,

No storm can shake my inmost calm,
While to that rock I'm clinging,
Since love is Lord in Heaven and earth,
How can I keep from singing?

The initial stage of my journey to find God brought me a wonderful sense of His peace and joy. My great hope is that others will take a journey similar to mine. Search for God. Find Him. Then, experience the incredible love, peace, and serenity only God can provide.

CHAPTER REFERENCES

PART I: TO BELIEVE OR NOT TO BELIEVE—THAT IS THE QUESTION

Chapter 2: Atheists' Views

Dawkins, Richard. 2009. *The God Delusion.* Houghton Mifflin Company Kindle Edition. (p. 51)

Chapter 4: The Implications of Atheism

Sinclair, D. A. (2019). *Lifespan: Why We Age and Why We Don't Have To.* New York: Atria Books. (pp. 119–139)

Harris, Sam. 2004. *The End of Faith.* W.W. Norton & Co. Kindle Edition (pp. 36–39).

Chapter 6: Evaluating the Case for Atheism

Dawkins, Richard. 2009 . *The God Delusion.* Houghton Mifflin Company Kindle Edition (p. 116, 28, 129).

Hitchens, Christopher. 2009. *God Is Not Great.* Twelve Kindle Edition (p. 373, 439).

Harris, Sam. 2004. *The End of Faith*. W.W. Norton & Co. Kindle Edition (p. 23, 15).

PART II: EVIDENCE FOR LIFE AFTER DEATH

Chapter 7: Life After Life

Janics Miner Holden, EdD., Bruce Greyson, MD., Debbie James, MSN, RN. 2009. *The Handbook of Near-Death Experiences: Thirty Years of Investigation*. Santa Barbara, CA: Prageger Publishers.
Kübler-Ross, Elisabeth. 1991. *On Life After Death*. Celestial Arts.
Moody, Raymond A. 1975. *Life After Life*. Harper Collins Publishers.

Chapter 8: Visits to a World Beyond

Sinatra, Tommy Rosa and Stephen. 2015. *Health Revelations from Heaven and Earth*. Rodale.
Alexander, Eben. 2012. *Proof of Heaven: A Neurosurgeon's Journey into the Afterlife*. New York London: Simon & Shuster.
https://www.hebrewcatholic.net/roy-schoeman/

Chapter 9: It's a Miracle

Keener, Craig S. 2011. *Miracles: The Credibility of the New Testament Accounts*. Grand Rapids: Baker Publishing Group.

PART III: RELIGIONS AND THE SEARCH FOR GOD

Chapter 11: Judaism, Christianity, and Islam

Haleem, M.A.S. Abdel. 2004. *The Qur'an*. Oxford University Press Kindle Edition.

Chapter 12: Discovering The Hebrew Bible

Friedman, Richard Elliot. 2017. *The Exodus*. Harper One Kindle Edition.

—. 1996. *Who Wrote the Bible?* Kindle Edition.

Lamsa, George M. 1933, 1961. *Holy Bible From the Ancient Eastern Text*. San Francisco: Harper Collins.

Nicolson, Adam. 2003. *God's Secretaries: The Making of the King James Bible*. New York, NY: Harper Collins Publishers.

Silberman, Israel Finkelstein and Neil Asher. 2001. *The Bible Unearthed*. Simon and Schuster Kindle Edition.

Chapter 13: Rediscovering the Hebrew Bible

Friedman, Richard Elliot. 2017. *The Exodus*. Harper One Kindle Edition.

—. 1996. *Who Wrote the Bible?* Kindle Edition.

Chapter 14: Science and the Hebrew Bible

Collins, F. S. (2006). *The Language of God*. New York: Free Press, p. 157.bbb

Archaeological evidence for Abraham: https://biblearchaeologyreport.com/2021/07/16/top-ten-discoveries-related-to-abraham/

DNA analysis showing relationship of Jews and Arabs to Abraham: (https://www.aish.com/ci/sam/48944786.html) (https://www.science.org/content/article/jews-and-arabs-share-recent-ancestry

Archeological evidence for Moses, Aaron, Passover and the Exodus: (https://www.thebereancall.org/content/dna-abrahams-children?sapurl=Lys5MjZkL2xiL2xpLyt3dm44dWs4P2JyYW5kaW5nPXRydWUmZW1iZWQ9dHJ1ZSZyZWNlbnRSb3V0ZT1hcHAud2ViLWFwcC5saWJyYXJ5Lmxpc3QmcmVjZW50Um91dGVTbHVnPSUyQnd2bjh1azg

Evidence for Cohen characteristics traced to Aaron: Watson, J. D. (2017). *DNA: The Story of the Genetic Revolution.* Alfred A. Knopf Kindle Edition.

Chapter 15: The New Testament

Blomberg, C. L. (2014). *The Historical Reliability of the Gospels: Second Edition.* Downers Grove, IL: IVP Academic: Intervarsity Press.

Nicolson, A. (2003). *God's Secretaries: The Making of the King James Bible.* New York, NY: Harper Collins Publishers.

Prager, D. (2018). *The Rational Bible: Exodus.* Regnery Faith Kindle Edition.

Tesoriero, R., & Han, L. (2013). *Unseen: The Evidence.* Kincumber New South Wales, Australia: Ron Tesoriero.

Chapter 16: Did Jesus Rise from the Dead?

Tesoriero, R., & Han, L. (2013). *Unseen: The Evidence.* Kincumber New South Wales, Australia: Ron Tesoriero, pp. 93–96.

PART IV: EVIDENCE FOR JESUS AS GOD

Chapter 17: Jerusalem, ca. 33 AD

Niyr, M. (2020). *The Turin Shroud: Physical Evidence of Life After Death?* Morgan Hill, California: Bookstand Publishing.

Wilson, I. (2000). *Jesus: The Evidence.* Washington, DC: Regnery Publishing.

Wilson, I. (2010). *The Shroud of Turin.* London: Bantam Books.

Chapter 18: Jesus' Burial Garments

Editors, C. R. (2014). *The Shroud of Turin History's Greatest Mysteries.* Charles River Editors.

Niyr, M. (2020). *The Turin Shroud: Physical Evidence of Life After Death?* Morgan Hill, California: Bookstand Publishing. (pp. 68–69).

Website: Shroud.com

Wilson, I. (2000). *Jesus: The Evidence.* Washington, DC: Regnery Publishing.

Wilson, I. (2010). *The Shroud of Turin.* London: Bantam Books.

Bennett, J. (2001). *Sacred Blood, Sacred Image: The Sudarium of Oviedo.* Littleton, Colorado: Libri de Hispaniea.

Chapter 19: Evidence from the Burial Garments

Niyr, M. (2020). *The Turin Shroud: Physical Evidence of Life After Death?* Morgan Hill, California: Bookstand Publishing.

Verschuuren, G. M. (2021). *A Catholic Scientist Champions the Shroud of Turin.* Manchester : Sophia Institute Press.

Chapter 20: Extremadura, Spain: 1326

Caso-Rosendi, C. (2017). *Guadalupe: A River of Light*. Front Royal, Virgina: First Light Press.

Chapter 21: Tepeyac Hill, Mexico: December 9, 1531

Chávez, E. (2006). *Our Lady of Guadalupe and Saint Juan Diego: The Historical Evidence*. London: Rowman and Littlefield.

https://dev-st-francis-the-americas.ws.asu.edu/sites/default/files/nican_mopohua_italian_nahuatl_spanish_english.pdf

https://www.catholiceducation.org/en/culture/catholic-contributions/saint-juan-diego-and-our-lady.html

Rosikon, G. G. (2016). *Guadalupe Mysteries: Deciphering the Code*. Ignatias Press.

Chapter 22: The Mystery of Our Lady of Guadalupe

Rosikon, G. G. (2016). *Guadalupe Mysteries: Deciphering the Code*. Ignatias Press.

Chapter 23: Fatima, Portugal: October 13, 1917

Bartold, M. (2014). Fatima: *The Signs and Secrets*. KIC.

Fatima: In Lucia's Own Words. (1996). KIC.

Marchi, F. J. (1947, republished 2015). *The True Story of Fatima*. KIC.

Chapter 24: The Children of Fatima

Fatima: In Lucia's Own Words. (1996). KIC.

Chapter 25: The Meaning of Fatima

Fatima: In Lucia's Own Words. (1996). KIC.

Marchi, F. J. (1947, republished 2015). *The True Story of Fatima.* KIC.

Chapter 26: Marian Apparitions

https://miraculousmedal.org/welcome/story-of-st-catherine/

Miraculousmedal.org

https://www.lourdes-france.org/en/bernadette-soubirous/

Chapter 27: Santa Fe, New Mexico: 1852

https://www.historicmysteries.com/loretto-chapel-staircase/

https://sspx.org/en/news-events/news/st-josephs-miraculous-staircase-2774

Chapter 28 Stigmata

Tesoriero, R. (2007). *Reason to Believe.* Ron Tesoriero.

Tesoriero, R., & Han, L. (2013). *Unseen: The Evidence.* Kincumber New South Wales, Australia: Ron Tesoriero.

Chapter 29: Signs from God

Tesoriero, R. (2007). *Reason to Believe.* Ron Tesoriero.

Tesoriero, R., & Han, L. (2013). *Unseen: The Evidence.* Kincumber New South Wales, Australia: Ron Tesoriero.

Chapter 30: The Blood of Christ

Cruz, J. C. (1986). *Eucharistic Miracles.* Charlotte, NC: TAN Books.

Eucharistic Miracle of Betania, Venezuela—December 8, 1991 (therealpresence.org)

Eucharistic Miracle of Sokółka, Poland—October 12, 2008—Part 1 (therealpresence.org)

(193) Doctor That Analyzed The Eucharistic Miracle of Tixtla—Interview Highlights (Includes Spanish)—YouTube

PART V: THE UNIVERSAL SEARCH FOR GOD

Chapter 32: God Loves Everyone

Cavins, J. (2015). *When You Suffer.* Cincinnati: Servant.

BIBLIOGRAPHY

Alexander, E. (2012). *Proof of Heaven: A Neurosurgeon's Journey into the Afterlife*. New York London: Simon & Schuster.

Alichieri, D. (n.d.). *Dante's Inferno: Translated by Rev. Henry Francis Cary: New Edition.*

Aquinas, S. T. (1989). *Summa Theologiae, Edited by Timothy McDermott*. Allen, Texas: Christian Classica.

Bartold, M. (2014). *Fatima: The Signs and Secrets*. KIC.

Bennett, J. (2001). *Sacred Blood, Sacred Image: The Sudarium of Oviedo*. Littleton, Colorado: Libri de Hispaniea.

Blomberg, C. L. (2014). *The Historical Reliability of the Gospels: Second Edition*. Downers Grove, IL: IVP Academic: Intervarsity Press.

Blot, F. R. (2016). *In Heaven We'll Meet Again*. Manchester, NH: Sophia Institute Pess.

Caso-Rosendi, C. (2017). *Guadalupe: A River of Light*. Front Royal, Virgina: First Light Press.

Cavins, J. (2015). *When You Suffer*. Cincinnati: Servant.

Chávez, E. (2006). *Our Lady of Guadalupe and Saint Juan Diego: The Historical Evidence*. London: Rowman and Littlefield.

Collins, F. S. (2006). *The Language of God*. New York: Free Press.

Cruz, J. C. (1986). *Eucharistic Miracles*. Charlotte, NC: TAN Books.

Dawkins, R. (2009). *The God Delusion*. Houghton Mifflin Company Kindle Edition.

Editors, C. R. (2014). *The Shroud of Turin History's Greatest Mysteries*. Charles River Editors.

Fatima: In Lucia's Own Words. (1996). KIC.

Friedman, R. E. (1996). *Who Wrote the Bible?* Kindle Edition.

Friedman, R. E. (2017). *The Exodus*. Harper One Kindle Edition.

Garza-Valdes, L. A. (1999). *The DNA of God?* New York: Doubleday.

Haffert, J. (1961). *Meet the Witnesses of the Miracle of the Sun*. Spring Grove, PA: Patricia M. Haffert.

Haleem, M. A. (2004). *The Qur'an*. Oxford University Press Kindle Edition.

Hamer, D. (2004). *The God Gene*. Doubleday Kindle Edition.

Han, R. T. (2013). *Unseen: the origin of life under the microscope*. Kincumber, New South Wales, Australia: Ron Tesoriero.

Harris, S. (2004). *The End of Faith.* W.W. Norton & Co. Kindle Edition.

Hawkins, D. R. (1995). *Power vs. Force.* Hay House, Inc.

Hawkins, D. R. (1995). *Power vs. Force: The Hidden Determinants of Human Behavior.* Arizona: Bard Press.

Hitchens, C. (2009). *God is Not Great.* Twelve Kindle Edition.

Janice Miner Holden, E. B. (2009). *The Handbook of Near-Death Experiences: Thirty Years of Investigation.* Santa Barbara, CA: Prageger Publishers.

Keener, C. S. (2011). *Miracles: The Credibility of the New Testament Accounts.* Grand Rapids: Baker Publishing Group.

Kelly, E. K. (2009). *Irreducible Mind: Toward a Psychology for the 21st Century.* Lantham, Maryland: Rowman & Littlefield.

Kübler-Ross, E. (1991). *On Life After Death.* Celestial Arts.

Lamsa, G. M. (1933, 1961). *Holy Bible From the Ancient Eastern Text.* San Francisco: Harper Collins.

Lewis, C. (1952). *Mere Christianity.* Harper One.

Marchi, F. J. (1947, republished 2015). *The True Story of Fatima.* KIC.

Marshall, T. R. (2009). *The Crucified Rabbi.* Dallas, TX: St. John Press.

Moody, R. A. (1975). *Life After Life.* HarperCollins Publishers.

Nicolson, A. (2003). *God's Secretaries: The Making of the King James Bible.* New York, NY: Harper Collins Publishers.

Niyr, M. (2020). *The Turin Shroud: Physical Evidence of Life After Death?* Morgan Hill, California: Bookstand Publishing.

Paine, T. (1794). *The Age of Reason*. Paris: Barrois.

Prager, D. (2018). *The Rational Bible: Exodus*. Regnery Faith Kindle Edition.

Rosikon, G. G. (2016). *Guadalupe Mysteries: Deciphering the Code*. Ignatias Press.

Ruffin, C. B. (1991). *Padre Pio: The True Story*. Huntington, Indiana: Our Sunday Visitor, Inc.

Silberman, I. F. (2001). *The Bible Unearthed*. Simon and Schuster Kindle Edition.

Sinatra, T. R. (2015). *Health Revelations from Heaven and Earth*. Rodale.

Sinclair, D. A. (2019). *Lifespan: Why We Age and Why We Don't Have To*. New York: Atria Books.

Tesoriero, R. (2007). *Reason to Believe*. Ron Tesoriero.

Tesoriero, R., & Han, L. (2013). *Unseen: The Evidence*. Kincumber New South Wales, Australia: Ron Tesoriero.

Verschuuren, G. M. (2021). *A Catholic Scientist Champions the Shroud of Turin*. Manchester : Sophia Institute Press.

Warren, D. M. (2020). *Brilliant*. Boston: Pauline Books and Media.

Watson, J. D. (2017). *DNA: The Story of the Genetic Revolution*. Alfred A. Knopf Kindle Edition.

Wilson, I. (2000). *Jesus: The Evidence*. Washington, DC: Regnery Publishing.

Wilson, I. (2010). *The Shroud of Turin*. London: Bantam Books.

ABOUT THE AUTHOR

R*obert Genetski* is one the nation's leading economists. He is a teacher, columnist, and author of 5 books on classical economic principles. His latest economics book, *Rich Nation, Poor Nation: Why Some Nations Prosper While Others Fail,* provides over a century of evidence for the success and failure of economic policies in the US and around the world.

For the past 25 years his consulting firm, *Classical Principles.com,* has provided economic and financial research to individuals and businesses around the world. Genetski is a popular speaker known for using humor and anecdotes to simplify complex economic issues.

Genetski taught economics at various institutions of higher learning including the University of Chicago's Graduate School of Business and New York University. He has held various positions in the financial industry,

Senior VP for a major Midwest bank, money manager, investment advisor and director of investment research, and has served as a Director on the Boards of various public and private companies.